IMPROVING RELIABILITY AND MAINTENANCE FROM WITHIN:

HOW TO BE AN EFFECTIVE INTERNAL CONSULTANT

STEPHEN J. THOMAS

INDUSTRIAL PRESS

Library of Congress Cataloging-in-Publication Data

Thomas, Stephen J.

Improving Reliability and Maintenance From Within: How to be
an Effective Internal Consultant

Stephen J. Thomas.

p. cm.

Includes bibliographical references and index.

ISBN 978-0-8311-3332-0

1. Organizational change--ment. I. Title.

HD58.8 .T41

658-dc21   2007

Industrial Press, Inc.
989 Avenue of the Americas
New York, NY 10018

First Edition, April 2007

Sponsoring Editor: John Carleo
Interior Text and Cover Design: Janet Romano
Developmental Editor: Robert Weinstein

10   9   8   7   6   5   4   3   2   1

This book is dedicated in
loving memory of my mother-in-law Arlene Silverstein.
Your presence enriched my life.

TABLE OF CONTENTS

Foreword

You're great—now change!

Most of what you read in the world of maintenance and reliability deals with strategies, processes, and technologies to master the elimination of the unexpected factory breakdowns.

Some of these strategies and processes such as Reliability Centered Maintenance (RCM) and PM Optimization (PMO) are proven effective when applied effectively. Technologies such as Vibration Analysis, Infrared Thermography, Airborne Ultrasound, Motor Analysis, and Oil Analysis have also proven to be very beneficial for achieving a reliable state in manufacturing or process environments.

Yet the industrial battlefield is littered with failed programs using any and all of the strategies, processes, and technologies mentioned above. These programs did not fail for lack of budget as management provided ample funds to achieve improved performance. They did not fail due to unqualified or untrained personnel. The personnel who attempted to implement reliability centered maintenance were fully trained and knew the techniques well. The staff that created the predictive maintenance and condition monitoring program were experienced and certified in the art.

Why are there so many casualties in terms of maintenance and reliability improvement programs that failed to live up to their potential?

They all overlooked one big thing—the culture of the workplace and the interest of the people who would be affected by these new programs.

Not many people are fully comfortable with change in most circumstances. Combine that with the fact that most maintenance and reliability professionals excel on the technical side, but may lack the requisite soft skills required to move the human condition toward change and you can see why some programs with great potential derail and fail to take root..

Now one among us, a true maintenance and reliability professional; has started a new discussion and has illuminated the path to navigate and transform people and change the workplace culture. Steve Thomas is the first and foremost in terms of the knowledge required to create a lasting cultural change—the kind of paradigm shift required if you plan to make big changes in your maintenance and reliability program.

Since 2004, we have had the honor of working with Steve to deliver educational workshops each year at the International Maintenance Conference, where 800–900 maintenance and reliability professionals seek new strategies, processes, and technologies. It is incredible to watch the students when they get to that "ah ha!" moment and all the reasons their past programs stumbled come into clear focus. Not only that, but they also leave with a clear path to making any new program much more likely to succeed.

I urge you to read this book word for word—and then order all of Steve's earlier books and workbooks—before you attempt any future improvement programs. It will not only move you to a different place at work. It will also move you to a different place in life.

Terrence O'Hanlon, CMRP
Reliabilityweb.com
http://www.reliabilityweb.com

PREFACE

I wrote my first book *Successfully Managing Change in Organizations: A Users Guide* to be a primer for those of us involved in the business of change. It was designed to be a tool to support those whose task it was to deliver value-added change to their organization. In that book I created the concept of the **Eight Elements of Change**. I also created the **Web of Change**® to represent the status of your change effort in a composite view with all of the eight elements represented on the same chart. In this book I dedicated two chapters to the **Eight Elements of Change** and provided the reader with a very brief description of each element.

After the first book was finished, I realized two things. First that there was much more that I wanted to say about the eight elements and, second, that there was another entire level that needed to be addressed if any change initiative was to be truly successful and long lasting. This second level was the organizational culture represented by the **Four Elements of Culture**. As a result I wrote a second book, Improving Maintenance and Reliability Through Cultural Change.

After about two years of making presentations, conducting workshops, and interacting with those in management who have been given the task of developing, deploying, and instituting organizational change, I once again recognized that something was missing. Although I had addressed the topic of change in great detail, I had missed one very important aspect: you and your needs related to successfully carrying out this task.

It dawned on me that all of us working through the change process and helping our organizations in this arena are actually internal consultants for our respective companies. We may be called other names such as project managers, special projects coordinators, team leaders, or other equally non-descriptive names. In the end, we are all internal consultants. As a result of this realization, I decided to write one more book in my Web of Change series, explaining how you can be a successful internal consultant within your company.

The result of this effort is in your hands. This book, *Improving reliability and Maintenance from Within: How to be an Effective Internal*

Consultant will explain what you have to do in order to develop, deploy, and assure continuity of any change initiative within your company. It will also help you and others understand both your true role and the value you can bring to your company when you are allowed to properly execute it.

This book does more than start you down the road towards being a successful internal consultant. It accompanies you on the journey. Enjoy the trip; it is worth it both organizationally and personally.

Acknowledgements

There are a great many people who deserve to be acknowledged for their help, support, and in many cases actual work performed to complete this book.

First of all, my wife Susan deserves a great deal of credit. As anyone who has ever written a book knows, it takes a lot of time. Susan's patience and her encouragement helped me immensely.

I would like to thank Peter Thomas and Amit Sabharwal for the design of the Internal Consultant's Web of Change. I know something about Excel, but the work that they did went far beyond anything I could have done myself. I would also like to thank Peter for the Business Case Model. It is a really good format that has many other applications beyond the one for which it was used in the book.

I would like to acknowledge John Carleo and Patrick Hansard of Industrial Press Inc. They are more to me than just my contacts with the publisher. They have been great advisors and supporters for my work and my ideas.

I would also like to thank Janet Romano of Industrial Press for her creativity in always coming up with a truly great cover. She is also the person who takes my Word documents and artwork and assembles them into an excellently formatted book.

My thanks are also extended Robert Weinstein, my editor. Robert is the person who takes my writing and fixes the problems so that you the reader can enjoy the content. He has a wonderful knack for fixing things, but never altering the concepts or ideas in the text.

I would also like to acknowledge Bob DiStefano and my friends at MRG, Inc., for showing me how a good external consultant works and how they can deliver value to their clients. In many ways, they have been my role models for the internal consultant work I perform.

I want to thank Terry O'Hanlon for writing the Foreword to this book and Reliabilityweb, Inc., for the support they have provided for my material. It has been a privilege for me to be able to present my change management workshop at their International Maintenance Conference year after year.

Last but not least, I want to acknowledge Norm King for his

ideas and his guidance, and for allowing me to become the internal consultant I was meant to be, and Joe Cresta for teaching me that people should always be treated with dignity and respect.

ABOUT THE AUTHOR

 Steve Thomas has more than 35 years of experience working in the petrochemical industry. During this time, through personal involvement at all levels of the work process, he has gained extensive experience in all phases of strategic development and implementation of organizational change. Together with a B. S. in Electrical Engineering (Drexel University) and two M. S. degrees in Systems Engineering and Organizational Dynamics (University of Pennsylvania), this broad range of experience has enabled Thomas to contribute significant value to the many projects he's worked on over the years. He has also presented workshops on successfully conducting organizational change for clients throughout the United States and Canada. In addition, Thomas has presented his change management workshop at many industry conferences that address improving reliability and maintenance – one prominent example being the annual International Maintenance Conference. While doing all this, he nonetheless has found the time and energy to write three books which constitute the *Web of Change Series.*

THE JOURNEY TO INTERNAL CONSULTING

Every journey begins with a single step.
For the internal consultant you need to make sure the first
and all subsequent steps are in the right direction.

1.1 My Journey to Internal Consulting

I never knew that when I started my career I was ultimately to become an internal consultant. In fact, I didn't even know at that time anything about internal consulting. In high school I had always thought that I wanted to be an engineer. I liked science and mathematics, working with and analyzing the detail these disciplines involved. At least that was what I thought. Based on that belief, I went to engineering school and became an electrical engineer. At the time I graduated, the economy was in a downturn with many layoffs and high unemployment. I went on many job interviews and actually got some offers to work in my profession with salaries that were in line with what other members of my graduation class were receiving.

Then one day I met a fellow student who I had not seen since I was a freshman. We had been on opposite work co-op assignments while in school so that when he was in school I was working and vise versa. During our discussion about what we had been doing for the last few years, he gave me the contact number for an employment agency which had gotten him a very good job, even better than what he could have achieved by interviewing though the school's in-placement service. I called the agency and they were the ones through which I ultimately got my first job.

The job that I got was far different than anything I had ever envisioned. I became a maintenance engineer in an oil refinery, learning about maintenance planning and work execution first hand by working on the front line. I really liked my job and never did go back to electrical engineering. That chance encounter with a fellow student set me on a path in life that I have found to be extremely challenging and rewarding—and a totally unexpected career.

Over the course of my career, I literally worked in every type of job in the refinery's reliability and maintenance organization all of the way up to assistant maintenance manager. I got to see and experience all phases of work and the work processes that supported the work. Then in the early 1990s, I was assigned to corporate headquarters to be the lead on a project that would install a common computer system for maintenance in all of the refineries in our company. Initially it all seemed so simple. Just rip out the old computer systems and replace them with the new one the project team had selected. This effort turned out to be far more difficult that what it appeared as on the surface. In fact, the effort took almost five years and had a major impact on my career and on my life.

One very important aspect of the project was the use of outside consultants. They served on two levels. First, they provided additional resources to help us with the vast amount of work that was required to move from five independent computer systems to one single common one. Second, the outside consultants counseled and worked with us to develop an improved maintenance work process that would ultimately be supported by the soon-to-be-implemented computer system. It was their contention that only by improving the work process and then bringing in the computer tool in support would we ever be able to achieve the improvements that we sought.

Their goal was to help us develop our improved maintenance work process first and then bring in the system as a support tool. To me this was very logical and made a great deal of sense. With this lofty goal, we held work process design sessions and got reluctant buy-in from the plant representatives.

At this time I was still learning about the concept of "soft skills," which I later developed into The Eight Elements of Change and The Four Elements of Culture. I also did not have a firm under-standing of the difficulty associated with change nor the various types of resistance that could be employed to undermine the change effort. Therefore, I incorrectly as-sumed that the tentative buy-in we had obtained for our work process improvement was sufficient for necessary implementation. This turned out to be very far from the truth.

Our next step was to present our project to the senior staff. Our premise for the project was that we should redesign the work process and then bring in the system as the support tool. However, the senior staff didn't see it that way. They did not really under-stand the need to improve the work process; nor did they appreci-ate the significant benefits that could be achieved as a result. They thought, as I originally had, that all we needed to do was to imple-ment the computer system and the rest would follow as if by magic.

As a result, our project team was literally kicked out of the sen-ior staff meeting only half way through our presentation. We were told to revise the approach and come back when we were ready to show how we were going to implement the software—nothing else. In fact, my manager at the time told me that in our next pres-entation I was not allowed to discuss "work process" at all.

So, as a result of our ill-fated initial meeting, we did what we were told and developed a project to replace the refinery comput-er systems with the one we had purchased as part of the project. Work process, structure, group learning, communications, interre-lationships, rewards, and most certainly leadership—seven of the eight elements of change—were essentially ignored and not addressed in the project's scope. The eighth element—technolo-gy—was covered because it addressed the replacement of the com-puter system; however, it existed in a vacuum with no recognition of the impact its installation would have on our business.

By forcing this direction, the management team essentially delayed by three years or more the benefits we should have deliv-ered to our company. In fact, several years later, when I was work-ing for one of the individuals who had been part of the senior staff the day we made our original presentation, I was told that in his opinion I had been right. Rather than feeling good about my vindi-cation, I really felt quite the opposite. I had not been able to convince the staff at the outset of the need to alter the work process and use the system as a support tool. The result was that I had not been able to help my company achieve the benefits sooner. Being right had no value to me, although a valuable lesson was learned.

Over the next several years, we installed the computer system at all of the plant sites. The conversion of old data, scheduling of the work around plant outages, and the required training all added time to the effort. During this period we discovered that the implementation of the system without process changes was causing great difficulty for the organization. The reason was that the system was based on sound work planning and scheduling practices, yet our sites were working in a far different manner—they were reactive in their approach to maintenance and did little or no planning. As a result, the system and those on the project were blamed for its inability to work in harmony with a process for which it was never created.

Also during this period, I enrolled in a graduate night school program to learn more about work process changes and other aspects of organizational dynamics. This interest was driven by my belief that improved work processes, among other things, were the answer to successful change management and improved plant reliability. Another reason that I enrolled in this program was my need to understand the process of change so that I could be of greater value to my company and to myself. Unbeknownst to me, I was taking the first step to becoming an internal consultant.

Just about the time that I finished this graduate program, the company went through a massive reorganization; I found myself demoted to the position of lead engineer handling capital projects at one of the plant sites. It was interesting work, but not the work that I truly wished to do nor where I felt I could add the most value to the company. It was also during this period that I became friendly with the site maintenance manager who shared my feelings about how we could improve plant reliability. Shortly thereafter, he instituted a massive redesign effort and asked if I could facilitate the process on his behalf. This was a very difficult task requiring a great deal of time and effort, but we were ultimately successful. As a result and a reward for my efforts, I was reassigned to his organization in a newly-created job—that of strategic projects coordinator.

This was a new and exciting role, one which had never existed in the maintenance organization. At the time, the organization was highly reactive; there were a great many strategic tasks that needed to be done if we were ever going to change from what we were to an organization that was reliability focused. It was my role to help develop and implement these initiatives because everyone else was preoccupied by the problems of the day. Although this may sound easy, it was far from it.

I knew that there were things requiring a strategic focus, but I was not completely sure how to go about approaching the work. As a result, the learning process that I went through involved a great deal of trial and error as I worked to define my new role. My learning experience was significant because I was able to see first hand the difficulty of the task. At the time I did not realize a more basic fact. I was doing far more than handling strategic work initiatives that the reactive organization was struggling to complete. I was also becoming an internal consultant, utilizing my experience and education to serve my company as well as my own career development.

1.2 Why I Wrote This Book

There are many people within reliability and maintenance organizations who do not have a job with specific duties that support the day-to-day operation of the business. These individuals are assigned to work outside of the day-to-day business process. They handle what are often termed special projects—the strategic projects that the organization needs to have completed, but never seem to get around to doing. These people often work alone or lead small groups of people assigned to work part time on these efforts. In addition, they are not always provided with the tools necessary to successfully accomplish the tasks to which they have been assigned.

This is not the best way for an organization to develop, roll out, and successfully implement initiatives that will ultimately provide them long-term benefits. The people occupying these positions are often looked down on by the balance of the organization because, after all, they do not appear to provide any day-to-day support of the business.

To further complicate matters, because they most often are not trained in how to develop and implement strategic initiatives, the organization has no buy-in to what is delivered. Various forms of resistance are usually the result. Even if the work effort is successful, the organization is often not aware of the difficult task that has been completed. As a result, the special projects team gets no recognition and hence no reinforcement to want to engage in this type of work in the future. They would much rather return to the day-to-day reactive work environment where they feel that they are part of the work team.

The bigger issue is that management, and even those who worked on the special project, may not see the larger significance of their work. They may only consider it as one more special project that they have completed. They have not been trained to see that projects of this nature contribute to a larger whole. These projects are initiatives that support the goal-setting process and ultimately contribute to achieving the organization's vision.

When it comes to those involved in special projects, everyone is missing the boat. The individuals involved in the effort and those who receive the benefits of the effort are totally unaware that long-lasting and beneficial change has been initiated by what I prefer to call internal consultants. There are several reasons for this. First, the organization is often so focused on the tactical nature of their work that they don't recognize the need for nor the significance of strategic projects. Second, there is no recognition by the organization that internal consulting is what is taking place when they execute strategic projects. To the organization, consulting is something that is handled by outsiders.

I wrote this book because those in special project roles and those who lead the organizations where individuals are assigned special projects need to have the proverbial light bulb go on; they need to recognize the value of special projects and those who handle them. If this can be achieved, then there will be a two-phased benefit for the organization.

First, those in charge will understand that you don't always have to hire a high-priced consultant to accomplish strategic initiatives. Often these skills exist right in the organization. They only need to be recognized and mobilized to gain significant benefit. When this takes place, those in leadership positions will enable the special projects people to take on a new role—that of internal consultant. They will see that there is more to the reliability and maintenance business than fixing what is broken. They will see that there are individuals in their organization who can see a strategic direction and positively influence the organization to move in that direction.

The other benefit is for those who handle the strategic projects. As I learned and they can learn as well, the role they play is often more significant than that of the planner or foreman tasked with fixing the broken pump or repairing the damaged piping in order to maintain production. They will understand that, although the planner and foreman have valuable jobs in the day-to-day work environment, they are the ones who are creating the future. They are the ones who are tasked with taking the organization's goals, developing initiatives, and delivering change to the company while at the same time getting buy-in and ownership from the workforce.

It took me a long time and required a lot of help by others for me to realize what I and others like me were contributing. However, when I recognized it, I can truly say it was a milestone in my development.

There is more to my rationale for writing this book than just helping organizations recognize the significance of the role of the internal consultant. Part of the problem is that once the organization and the individuals working in this capacity recognize that significant contributions can be made in this manner, they begin to try to figure out what specifically an internal consultant does and what tools are needed so that they can reap the benefits.

There are a great many skills and tools that you need for the work of internal consulting. While it is true that you can acquire the tools individually, there is nothing out there that can give you the whole tool set. I know this to be true because I have looked. The reason for this is that the majority of writers are focused and skilled at providing information about the individual tools such as facilitation, coordination, project development, goal setting, and others. Part of the reason for this is that academically each of these topics can easily be a book unto itself. Although learning the detail of any one of these topics is important, it is secondary to learning and applying an entire tool set to the role of internal consulting.

The other part of the equation is that internal consulting skills can really only be taught by someone who has done the work—someone who has been an internal consultant and experienced first hand the associated problems and pains that accompany the job. The books I have read about consulting are written by successful external consultants. An external consultant plays a role that has some similarities, but ultimately has great differences as well. External consultants don't have to continue working in the plant among their peers after their engagement. External consultants don't have to overcome credibility issues by an organization that often doesn't view special projects as very important. For these and other reasons, a book for internal consultants should be by someone who has been one.

In my case I have "been there and done that." My recognition of who and what I truly am, as well as my success in building not only a special projects role but a strategic projects (internal consulting) role, will enable me to explain it to others so that they can achieve the same status without the same amount of struggle.

1.3 Who Can Benefit

This book, focused on how to become a successful internal consulting, can provide benefits at many levels.

Special Project Coordinators

For those out there who fit this role, there is more to the job you are performing than you realize. You have been assigned the special project on which you are working, probably because your management thinks that you have the experience and skill needed to carry it out. If you have worked with external consultants, you will recognize that these are some of the tools that they also bring to the work effort. What you need to recognize is that you are also doing consulting, except it is from the inside of the organization. While you may not have had the advantage of working at many different companies, you have something the external consultant does not. You have detailed specific knowledge of your plant and the people in it. You know what makes them tick and how to get work done. You can benefit from this book because it will provide you with a set of internal consulting tools that will enable you to do your work better and add long-term value to the business.

Emerging Internal Consultants

Once special projects coordinators recognize that they are actually internal consultants, there will arise the need to clearly define this new role not just for themselves but also for their organizations. In doing this, they will begin to see how the project that they are working on has ties to and supports other strategic projects within the organization. They will begin to see how the full set of strategic projects fits together into a larger whole. This recognition allows them to begin to collaborate with others to achieve the larger and more far-reaching benefits. This in turn enables them to acquire additional skills and more experience, and will ultimately lead to improved work.

Seasoned Internal Consultants

As internal consultants become seasoned, they begin to recognize not only how other projects fit together, but also the gaps that are not yet addressed. Because they need champions and sponsors within the organization to initiate the work to close these gaps, seasoned internal consultants learn how to acquire these sponsors and champions. They learn to promote initiatives that close the as-yet unrecognized gaps in the work process. When this stage is reached, the internal consultants recognize this ability within themselves; furthermore, the organization recognizes it and utilizes it as well.

External Consultants

This book also has benefit for external consultants. Many consultants come into a company to perform work and either dismiss or, at best, provide cursory recognition of the internal consultants. This is a mistake because, if the external consultants can learn to work in a partnership with those working inside the company, there is much to be gained. The external consultants can gain great organizational insight, making their job easier. The internal consultants can acquire additional skills because the external consultants employ a developed consulting discipline. This discipline can be learned and then employed later, after the external consultants are gone. The internal consultants can also gain insight into how the work is handled by other companies within and outside of their industry because external consultants bring this knowledge to the table.

Organizational Leaders

This book can benefit the leadership in the organization by helping them recognize the untapped value of their internal consultants. Once the true role of the people who they have assigned to "special projects" is recognized, they can enable them to deliver the value to the organization that their experience can provide. If developed correctly, the organization's leadership will have created a strategic projects team that can be continually focusing on the future. Because most organizations, hampered by the day-to-day problems, are struggling to become reliability focused, the role of the internal consultant can be extremely important.

Organizations in General

Typically organizations look down on those in special project roles. After all, they are not actually part of the day-to-day work. Once the organization recognizes the true value of internal consulting, they will be able to tap into it. This will help them accomplish those projects that they never seem to have time to do in a way that will focus them on future improvement.

1.4 Why Is This Book Different?

This book is very different because a book of this type doesn't exist anywhere else. While there are books on individual aspects of consulting designed to provide information about a specific topic there are few if any that treat the subject as a complex whole. Further to the best of my knowledge there are no books out there in the market that recognize individuals within companies not as simply being assigned special projects but rather as the role they really are playing—that of internal consultant for their firm. The significance of this is two-fold. First is that the book provides recognition on a personal level so that those in the special project roles can recognize that they are far more important than even they realize. The second significant aspect of the book is for company recognition. Often organizations and their leaders treat the special projects people as outcasts since that are not linked to the day-to-day and do not directly support production. This book helps the leadership within these companies recognize that these individuals can add far more value than they typically are permitted to deliver. Once this hurdle is overcome companies will enable their internal consultants to add value and benefit as a result. This is an entirely different view of special project work. It no longer is something that is additional work; it is work that is being developed so that the organization will have a new and more productive future.

1.5 What Is Included?

The book is seventeen chapters in length. The initial chapters (1–6) define an internal consultant. They provide the business case as well as other significant components of this rather complex job. The next section (Chapters 7–13) discusses clients and the details about the work. It is important that internal consultants understand how to obtain and keep clients and how to deliver a valuable work product. This section is followed by a detailed description of how projects are brought to completion as well as how you can build sustainability into the work (Chapters 14–15). Each of these chapters has a set of five questions at the end. The goal of these questions is to get you thinking about the subject matter of the chapter and how you, the internal consultant, can put this information to use.

The end of the book (Chapter 16) focuses on you the internal consultant and how you can go about maximizing your effectiveness in the internal consultant's role. Included in this chapter is the introduction to the Internal Consultant Web—a 12-spoke web diagram that will help you assess your strengths as well as the areas where you can improve. The web and the actual web survey are included on a CD and in Appendix 3 at the end of the book.

Chapter 17 wraps up the discussion of internal consulting. It offers encouragement as you begin a journey that has the potential of immense rewards both for your company and for you on a far more personal and developmental level. A more detailed description of the chapters follows.

Chapter 1: The Journey to Internal Consulting

This chapter introduces the concept of an internal consultant and explains the purpose behind this book. It is rare that people working on special projects in their organizations are thought of as internal consultants. Although they most often are doing strategic work, people who work on special projects are never recognized as being consultants in their own right. This chapter starts the reader down the path of self discovery if they are performing this work. It also helps to increase awareness by the organization's leadership of the value of the internal consultants within their company.

Chapter 2: What Is an Internal Consultant
In Chapter 2 we define in depth what an internal consultant is and what it is that they do. This includes the roles and responsibilities of the position. This information will help define the role as well as identify areas where those in this role can add additional value to their position. It will also clearly delineate the traits and skill set required of someone who aspires to this position.

Chapter 3: Strategic vs. Tactical Work
Once it is understood that internal consultants are needed and can add value, we need to develop a process to enable them to do their job. Just as with external consultants, those working internally can not do the job in a vacuum. They need the efforts of those in the organization that are almost always consumed with the day-to-day efforts of keeping the plant running. This chapter describes the difference between strategic and tactical work. It also explains in detail how to engage the tactical workforce in strategic work initiatives.

Chapter 4: The Business Case for Internal Consulting
There is a clear need for people within an organization that possess detailed reliability and maintenance knowledge to work with the organization as internal consultants. Without people filling this role, the organization will not advance. It is impossible to handle both strategic and tactical work at the same time. This work is accomplished in different hemispheres of the brain and they do not communicate with each other very well. In addition, the majority of those working in the reliability / maintenance arena simply don't have the time to conduct projects that involve longer-term strategic efforts. This is the reason and the need for the internal consultant. Many companies exist to provide external support for companies who have needs in the strategic area. However, this

need can also be filled by internal consultants who bring a different set of skills to the work effort. This chapter makes the business case for establishing internal consultants within your company.

Chapter 5: Organizational Culture and Soft Skills

Part of being able to successfully support a change initiative is the understanding that simply implementing a new work process is not the path to success. Beyond this level of change referred to as "hard skills" are two other levels that are even more important. The first is referred to as "soft skills." This level is made up of eight key elements called the Eight Elements of Change. The other level addresses the organization's culture. It is made up of four critical elements called the Four Elements of Culture. Chapter 5 discusses these additional levels in support of the internal consultant working to deliver successful change.

Chapter 6: Learning Organizations

For organizations to effectively utilize the output of the consulting effort, they need to be learning organizations. They need to be able to set goals and initiatives, conduct the work, and evaluate the outcome vs. the desired end state. When these are out of alignment, as they often are, learning organizations will be able to develop corrective action plans. They then feed this back into the ongoing activities to bring them closer in alignment with their goals. In addition, true learning organizations need to evaluate their goals; if the goals are not correct, they need to modify them as well. This chapter explains how this can be accomplished and, more specifically, how the internal consultant can help make this happen.

Chapter 7: Clients

As an internal consultant, you may have excellent ideas about how you can help the organization make a change for the better. That is all well and good. However, without clients, none of your ideas will ever reach fruition. That is simply because, as an internal consultant, you are there to enable change, not own or lead it. The ownership and achievement of the goals are left to your clients. Therefore, you need to understand how to get and retain clients in order to help them with their initiatives. This understanding will let you develop supporting initiatives of your own. It will also help you enable your clients to deploy them.

Chapter 8: The Work Process
Once you have clients, the next step is defining and doing the work. Your initial work effort will be the result of a client asking you to handle a project. However, this doesn't get the internal consultant what they really need because it is most often a one-time event. This chapter will help you to get past the initial assignment. It will help you build your credibility not only to get additional work, but also to be included in the detailed development of the other work initiatives. It will also provide insight in how to build your credibility so that you will continue to be engaged by your company in additional initiatives.

Chapter 9: Resistance
Every change initiative is accompanied by some form of resistance. This is not something to be overcome. Instead, it is evidence of the organization's need to learn how to adopt and adapt to new ways of doing things. The information provided in this chapter will help the internal consultant identify and accommodate resistance as part of the change process.

Chapter 10: The Internal Consultant's Role
This chapter deals with all of the skills needed by the internal consultant to handle the initiatives in which they become involved. The work they do in this area has some similarities and some major difference from the skills employed by external consultants. An internal consultant often has far more organizational knowledge than do those hired from outside. As a result, they can facilitate and coordinate initiatives in a different manner. Whereas external consultants often focus all of their efforts on the job at hand, a good internal consultant can link the project to others that have or have not yet started. By doing this they can add additional value to the effort and create additional work for themselves. In the area of consensus and conflict resolution, they also play a different role than the external consultants. The main reason is that they know the people involved – often very well and they have to survive after the current initiative has been completed.

Chapter 11: Work Teams

One of the most significant jobs of the internal consultant is mobilizing and working with teams. These can be day-to-day work teams or teams assigned to work on a short or long duration project. In any case, getting the team organized and positioned to deliver the sought after work product is important. Another aspect of this topic is one of ownership. Internal consultants can not own the work or the work product. If they do, the end result of the initiative will be failure. What is needed and delivered by this chapter is how to build ownership so that, once the project is completed, the organization owns the result.

Chapter 12: Working with External Consultants

As good as internal consultants are, there are numerous times when external consultants can bring value to an initiative. However, companies that place all of their eggs in the external consultant's basket are missing a vital component. Although these initiatives may require an external view, blending what they have to offer with the organizational savvy of the internal consultants can add even greater value and bring extraordinary results. This chapter focuses on how to blend the skill sets of these two similar yet diverse groups.

Chapter 13: Business Ethics for Internal Consultants

Ethics essentially dictate the correct way to behave. Ethics are a very important part of internal consulting because the consultant often is working in the difficult environment of organizational change. The correct behavior establishes the consultant as someone with high standards and credibility—someone who can be trusted to work with the organization to deliver the value they expect from the changes that they are trying to put in place. Improper ethical standards for the internal consultant's work leave them without this level of trust and make working in the consultant's role virtually impossible.

Chapter 14: Completion of the Work

Every initiative reaches completion at some point in time. Typically work process initiatives do not really end because, as you complete one phase, you evolve to the next. Nevertheless, phases of work process and other related changes do end. This chapter discusses how to complete one of these stages so that it can be assimilated into the organization's culture and, as a result, deliver long-lasting benefit. This chapter also addresses various ways in which to complete initiatives so that the client learns and continues to progress to the next phase.

Chapter 15: Readiness and Sustainability
This chapter addresses two key points if an initiative is going to be successful: readiness and sustainability. Readiness focuses on the site personnel and the work that needs to be done to get them ready to make a change. This work needs to take place at the beginning of the effort so that what is developed will be accepted. Sustainability takes place after deployment. It addresses the requirements and actions that need to be taken if you wish the initiative to survive. If sustainability is not addressed, you may be successful in the short term, but what you have developed may not last very long. Because an internal consultant doesn't actually leave the site, failure from lack of readiness or a plan to sustain the effort is harmful to your credibility and to your ability to continue doing consulting work.

Chapter 16: Improving Your Internal Consulting Skills
Becoming an internal consultant is easy; becoming a really good one takes a lot of hard work. Internal consultants need feedback on the work that they deliver to their organization. This is easier to acquire if you are an external consultant because the senior members of your company will invariably be working with you on the engagement and will offer you constructive criticism as an integral part of the work engagement. The internal consultant does not always have this opportunity.

Apart from the standard yearly performance review, there is usually little feedback provided related to an individual's work performance. The internal consultant needs more timely and directed feedback. They have two avenues to obtain this feedback. They can perform a self assessment or they can ask for feedback from their clients. This chapter addresses how to do this and what to do with the results. Also included is the internal consultant's web of change—a tool to help make this process a success.

Chapter 17: The End of the Beginning

Chapter 17 summarizes the material presented in the book and closes out the internal consulting discussion.

1.6 Navigating Through the Book

My suggestion is to read the book from beginning to end. I say this because the majority of the chapters are sequential in nature and they build on each other. However, there may be some out there who have very specific interests and would like to read through the chapters out of order. To enable this, I have provided the diagram in Figure 1-1. It shows the chapters and how a reader could review them out of order.

Essentially there are three tracks. The first is covered in Chapters 2 through 6. These chapters lay out the reasoning behind the internal consulting process. The second track covered by Chapters 7, 8, 9, 10, 11, 12, 14 and 15 address the internal consulting process. A reader who already understands the reasoning for internal consulting may follow this track to obtain a better understanding of the process. The last track, covered by Chapters 13 and 16, addresses several special topics. Although these topics are part of internal consulting work, they can still be reviewed and deliver value in a stand-alone fashion.

1.7 Getting Started: A Message to the Reader

What you are about to embark on is not going to be an easy path to follow. Internal consulting is not a recognized skill in the reliability and maintenance arena. Most often, the role you a filling is still looked upon as that of a special projects coordinator. In the eyes of those doing the day-to-day work, the role may not even be considered a vital part of the work effort. Your battle for self recognition and of recognition by your organization is going to be very difficult. However, if you stick with it, you will ultimately be able to deliver very significant value for yourself and your organization. As you work at the task of becoming a successful internal

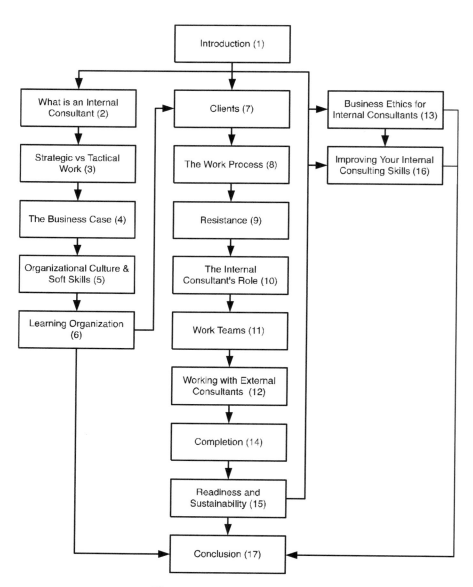

Figure 1-1 Book Navigation

consultant, consider what Machiavelli the Italian statesman had to say about change over 500 years ago in The Prince:

"There is nothing more difficult to take in hand, more perilous to conduct, or more uncertain in its success than to take the lead in the introduction of a new order of things, because the innovator has for enemies, all of those who have done well under the old conditions, and luke-warm defenders in those who will do well under the new."

Five Things to Think About or Do

1. Do you have people in your organization who handle "special projects?" Are they treated as important members of the organization?

2. Do those who handle special projects do this work as a part-time effort? If so, how successful do you think they are in accomplishing the special project tasks?

3. Does your management team fully recognize the value of longer-term strategic efforts and those who handle them? If they do, name several strategic initiatives that have benefited the organization. If they don't, list a few that would help the organization improve.

4. Do you have recognized internal consultants in your organization? If so, have you worked with any of them? Can you see the value they offer to the organization? Take the time to sit down with one of these people and learn more about their role in the company.

5. The navigation diagram (see Figure 1-1) can help you select specific topics that may help you perform your work. Please send any questions to Changemgt@gmail.com.

What Is an Internal Consultant

Support from within to affect positive change
As a basic business concept shouldn't be strange

2.1 External Consulting

Most of us have had a stereotypical experience with an external reliability or maintenance consultant. Our management hires a consulting firm that comes in to the plant with the intent of helping us to improve in areas where our management feels there is a deficiency. The consulting firm is often one we have heard of, met at one of the various reliability / maintenance conferences, or read about in the trade magazines. They are the recognized experts.

The first step is usually the introduction of the consultants by management. This introduction is often handled at the manager's staff meeting for the senior people in the organization and at group meetings for the balance of the organization. These meetings involve more than simple introductions. The senior management also uses these "get to know you" meetings as a way to explain why the consultants have been retained and what they hope will be the outcome of the consultants' work. They hope that by explaining the purpose—whether it is to improve maintenance planning and scheduling, institute preventive maintenance programs, or to improve workforce productivity—they can get the organization and others involved to understand their rationale. After all, consultants working on a change initiative are a disruption to the normal day-to-day work; the support of the organization is needed to make the work effort a success.

The second reason that these introductory meetings are held is because the majority of those seeing consultants on site have a nagging fear that there is some ulterior motive behind the engagement. Is the plant going to be shut down? Is there going to be a major restructuring or layoff? Or will there be some other outcome that could very negatively impact the workers and, ultimately, their livelihood? The introductory meeting is designed to dispel these negative ideas. Often this works, but often it is just more fodder for the gossips and additional concern for the workforce.

The next step is the information gathering process. In this step, the consultant team (usually there is more than one person) goes into the plant and interviews people across the organization whom management believes 1) are involved with the issue that they are trying to correct, and 2) will be the most open and honest in describing the problem. Depending on the size of the problem being addressed and the availability of those being interviewed, this process can take a considerable amount of time. However, in the end, good consultants will have developed a very clear picture of the problem. They will also have some very clear ideas based on their experience of things they can do to help correct it.

Next the consultants adjourn to the office that they have been provided on site, or they leave and go back to their office to develop their report. In this report, they summarize what they have learned. They make recommendations regarding what they believe the organization needs to do to achieve the desired improvement. When this report is completed, it is presented to the management team that hired the consultants to do the work. These presentations often take a considerable amount of time. There usually are a great many questions asked as the management team attempts to get a better understanding of the consultants' understanding of the problem and the suggested corrective action plan.

The report presentation also has two additional components: 1) the consultants' plan of action if they were allowed to work with the client company to implement the change, and 2) the cost of the consultant services.

At this point, management needs to make several decisions.

Do they believe what the consultants are telling them about what needs improvement and how it needs to be done?

Some companies immediately recognize that the consultant findings are exactly correct and they agree on the course of action recommended. Others, while they see the need to make the changes recommended, often don't agree with the strategy to go about it. Still others who never really wanted the consultant in the first place will get defensive and dismiss the report and the findings as invalid.

Are they ready to proceed with the change?

Assuming that the company believes that the recommendations are valid, are they really willing to commit to do them? Change of the nature we are discussing is not simple, nor something that is done overnight. It requires a great deal of hard work by the entire organization over an extended period of time. The management of the company needs to make this commitment if the recommendations of the consultant are to be achieved.

How do they want to proceed?

Assuming that the company wants to take on the challenge, there are several ways that the work can be done. First they can retain the consultants to do the work. This is not a recommended approach because, if the consultants do the work, there is no on-site ownership. Good consultants will try to persuade against this approach; many will refuse to work in this manner knowing full well the results will not be positive.

A second preferred approach is for the consultants to work with the organization, providing coaching and guidance throughout the effort. Certainly this is what the consultants want because this type of work is the heart of their business and has a high degree of potential success, due to site ownership. This is also why the final report not only has a work plan but also has the consultants' cost to work in this fashion throughout the duration of the change initiative.

An alternative is to use a mix of the consultants' organization and of the site organization to do the work. In this manner, the company can reduce the cost of the effort and involve their own personnel, providing site buy-in to the effort. Last, and not recommended, is to thank the consultants for their effort and take on the work solely with site personnel using the consultants' recommendations. The client company's lack of experience in change initiatives is the main reason this is not a recommended approach.

I have witnessed each of these scenarios unfold many times over the course of my career. The companies that dismiss the information as invalid, or worse, get defensive about the information, have just wasted a great deal of time—not just their own, but the time of the consultants as well. Those companies that will agree the information is valid, but claim lack of readiness to take on the work have done the same thing, but in a different manner. The former shows active resistance to change while the latter show passive resistance. Nevertheless, in either case, the work does not get done and the potential value of the initiative is lost.

Those that decide to make the change often do not go to the extreme of trying to accomplish the task alone. They recognize that they need to retain the consultants for their expertise and objectivity for some period of time. The time the consultants are on site can range from just helping to get the effort started all the way to being on site until the initiative is completed. There have even been extreme cases where the consultants are on site for years as the change is instituted and ingrained within the organization.

Nevertheless, there comes a point where the work scope assigned to the consultants ends and they leave to move on to the next engagement. The point is that for external consultants there is always an end point. That is one of many significant differences between internal and external consultants. External consultants, as we have discussed, do the work and leave; internal consultants are different. Although the process is somewhat the same, they have to live with what has been created!

2.2 Internal Consulting

Whether they recognize it or not, every reliability and maintenance organization has at least one or possibly more internal consultants. These are the individuals who invariably handle the "special projects" as either a full-time or part-time job assignment. They have a great deal of practical and organizational experience. They have knowledge about the business and its people as well as a track record of getting difficult multi-functional tasks completed. They are the people who management at all levels often relies upon for guidance when new initiatives are proposed.

In our example above, the site's internal consultants would have been the ones who coordinated the external consultant data gathering / reporting effort. They probably would also be the ones who would lead, or at least support, whatever the company decided to do with the results of the consultant's report.

Internal consultants can also be utilized to fill the role of external consultants. In our example, if the company did not want to bring in an external consultant to conduct the analysis and make recommendations, a good internal consultant could be used.

At one point in my career, I was asked to conduct an assessment of a maintenance organization within one of our plants. The maintenance manager, while receiving favorable monthly metrics, had a suspicion that things were not as they appeared in the reports he was receiving. In my role of internal consultant, I essentially performed the same tasks as an external consultant. I ran the introductory meeting, conducted extensive interviews, and prepared and presented a report to the managers based on my findings. The results of the assessment were what the manger had expected. Although the metrics looked good, the work process and what was actually being done in the field were far from what was being reported in the measurements.

Just as the external consultant would have done, I also prepared a report of my findings and what I believed were recommendations for corrective action. The difference was that I needed to be very careful in what and how I presented my findings. Many of the people I had interviewed were my peers; I was going to have to work with them on future efforts. Needless to say, I spent a considerable amount of time crafting the report. I wanted to clearly point out the problems, make sound recommendations for corrective action, and (most important!) survive to work another day—a major differentiating factor between internal and external consultants.

2.3 Consulting Defined

Consulting in its various forms takes place throughout our business all of the time. It may be supplied from external sources or internally by those who work on special projects and are considered internal consultants to the business. In either case, consultation techniques are applied in order to take an objective look at how things are done and provide possible alternatives to make them better. This is certainly true in the reliability / maintenance arena where we are constantly striving to improve in order to move from reactive to proactive, reliability-focused maintenance.

To be able to understand what consulting in the reliability / maintenance arena is all about, we need a comprehensive definition describing this function. In this way, those working in this area as internal consultants—knowingly or unknowingly—can clearly define their role in the grand scheme of things.

Consulting, whether internal or externally applied, can be defined as:

> The ability to apply broad-based knowledge and experience about a specific area of business to help develop and implement strategic improvement plans, identify performance gaps, develop and support the implementation of a recommended plan of action to close the gaps, and provide the tools for long term sustainability of the initiative's deliverables.

In order to further clarify this definition, let's break it down and discuss the component parts.

Applying broad-based knowledge and experience

It is virtually impossible to work in a consulting role without broad-based knowledge of the type of work you are trying to improve, and without experience. For external consultants, this is obtained by working for numerous client initiatives across a wide variety of companies. In fact, consultant firms design the staffing assignments of their consultants in such a way as to provide this experience to them. They can then apply what they have learned to address the needs of the client at hand.

Internal consultants have a far more difficult time acquiring this knowledge and experience. While it is true that many have learned and gained experience working in other companies, it is not usually as broad-based as the external consultant. Their prior positions were usually at lower levels in the organization as they gained experience preparing for their current job. In addition, they most likely have not worked at a great many firms when compared to the experience gained by their external counterparts.

The same sort of problem exists for internal consultants who have worked for the same firm for many years. Although they may have held numerous positions, the work experience is still limiting. These reasons explain why internal consultants are a rare breed. Broad-based knowledge and experience for the internal consultant is hard, but not impossible to obtain.

About a specific area of business

Because of the complexity of the task, reliability / maintenance internal consultants typically work in a specific area of the business. The department in which they work is usually the one into which they are hired. Although people do change jobs as a result of increased experience, improved performance, and promotion, they are less likely to change departments.

People hired as reliability engineers, inspectors, maintenance planners, etc., have a high likelihood of staying in this type of work. The reasoning is that, as they gain experience in a specific area, they become valued assets and very difficult to replace. Even when they occasionally do change departments, they bring their learning and experience with them and apply it to the new position. For example, maintenance professionals who are transferred to Production bring along their maintenance perspective and are able to apply it to the new job. Additionally, this job change across departmental boundaries is often developmental in nature. After a period of time in order to gain some diversified experience, they revert back to their original group.

Develop and implement strategic improvement plans

Consultants need to help deliver strategically-focused improvement plans. The key word here is strategic. To help a company develop and improve their performance, strategic plans—as opposed to the tactical day-to-day plans—are required.

Consultants need to be able to work consistently at the strategic level in order to be able to deliver this value. This is far easier for external consultants than internal consultants to accomplish. Internal consultants tend to be given the special projects that are usually strategic and long term in nature. However, internal consultants are also often required to work on the current, tactical day-to-day work efforts of the organization.

For example, implementing a preventive maintenance program where one never existed before is strategic because it addresses developing a work process far different than the current reactive mode of operation. The problem faced by internal consultants is that they often are handed this strategic effort and expected to do this work along with their normal day-to-day tactical assignments. It doesn't work. The separation of strategic and tactical work will be addressed in Chapter 3.

Identify performance gaps

One of the key roles of consulting is to help the client identify performance gaps. These are the areas for improvement that, if closed, will enable the site to move closer to the reliability / maintenance vision they have created for themselves. This is a critical part of consulting. However, identification of performance gaps is not always a simple task. Consider an organization that wishes to implement a preventive / predictive maintenance work process where one has not existed in the past. On the surface it would appear all that was needed was to develop the program and deploy it across the site. Not so! There are three levels associated with identification and closure of performance gaps if a company wishes to achieve a successful outcome. These are:

Hard Skills

This level corresponds to the actual task that needs to be accomplished, such as the deployment of the preventive maintenance program.

Soft Skills

This is the foundational level where elements such as leadership, work process, structure, group learning, technology, communication, interrelationships, and rewards are employed to directly support the hard skill initiatives.

The Organizational Culture

This level is sub-foundational, where changes in the culture of the organization are addressed, enabling the soft skill and ultimately the hard skill changes to be successful.

Develop and support the implementation of a recommended plan of action

Once the gaps are identified, the consultants apply their knowledge and experience to develop a gap closure strategy. This may be done independently, but a far better approach is to develop the strategy with involvement of those on site who will be the ones to make it successful. This will also go a long way to ensuring long-term sustainability of the effort. Developing gap closure plans is not an end unto itself. The plans must be communicated and implemented at the site level in ways that will cause them to stick.

This is the hard part for any consultant and even more difficult for those working internal to the organization. While external consultants have pride in their work and a great desire to leave behind value for the organization, the internal consultants are not leaving. Therefore, they have a far greater stake in the outcome. For this reason, they tend to get more deeply involved in the actual work. This can be very helpful at the inception of the effort, but it is disastrous for long-term sustainability. That is why part of the definition specifically says "support" the implementation. Those who actually do the work must be those who will own it into the future—the site personnel. Without this level of involvement, the initiative will not survive.

Provide tools for long term sustainability

In the reliability / maintenance arena we are all engineers or, if we have advanced through the ranks, we have been taught to think like engineers. To me this means that we have been taught to think of all the things that we do as projects—work efforts with a beginning and a finite end point at which time the work is completed. Change initiatives are not projects! They certainly have a beginning, but they do not have an end. Change initiatives are efforts that can only be associated with continuous improvement; they have no end point.

This concept is understood by the consultants, but generally not by the site personnel involved. Therefore, it is the consultants' responsibility to get people to understand this rather alien concept. Not only that, but they also need to work with the site personnel to put in place procedures, plans, practices, long-term work teams, and other strategies that assure the work product remain viable and continues to grow. This is even a larger issue for internal consultants. Because they remain on site, it often falls to them to provide this continuous support as the other team members return to their tactical work. This is a trap that must be avoided, otherwise, the site will lose ownership and the initiative will fail.

Consulting is a task. Those who actually do the work are consultants. The definition of consultants describes their abilities and can be applied to people working inside the client organization, or those hired from a consultant firm to help perform the work. A consultant is defined as:

> A consultant is an individual who has the ability and skill to perform consulting work in a way that optimizes site ownership so that, when the initiative is completed, the site personnel have ownership of the outcome now and into the future.

In our definition of consulting, we described what this type of work entails. The definition of consultant is far simpler. The key aspect of this type of work, beyond employing their knowledge and skill to identify and close performance gaps, is the ability of the consultant to promote ownership. Without ownership of the end product, all is lost. The motto for the consultant should be:

> "At the end of my work on the initiative, regardless of the time and effort expended, the site personnel should own the result; they should believe that they accomplished the task on their own with a little consulting support."

2.4 Internal Consulting Traits and Skills

Not everyone, even those assigned to special projects, has the ability to be a good internal consultant. There are three elements that are required for success: position, personal traits, and internal consulting skills. The first is provided to the individual, the second is part of the individual's make-up, and the third is learned.

Position

Position is the placement of the individual within the organization's structure. People at the bottom of the organization will not have the kinds of peer level associations that will enable them to be effective. Those at the top are the leaders; as a result, they can not act in a consultant role within their own organization. The best position to be in is on the manager's staff in a solely strategic role. In that way, the internal consultants can obtain leadership support yet interact with their peers. Positional qualities include:

Working within the department making the change
Working at a peer level of those on the manager's staff
Strategically focused (little or no day-to-day work
 involvement)
Supported by multiple champions within the
 organization (referential power)

Traits

Traits are distinguishing personal qualities of an individual. They enable some to be effective field execution leaders and others to be good internal consultants. Internal consultant traits include:

Credibility
Passion for the work
Honesty
Ability to achieve one's ends through influence
Open to new ideas
Ability to see the big picture (the forest for the trees)
Good communicator
Ability to let others take the lead and the credit
The desire to make improvement while doing no harm

Skills

Skills, especially those related to internal consultants, are learned abilities. They enable us to use our knowledge to effectively and efficiently support organizational improvement initiatives. For internal consultants, the skills needed include:

Experience in the business line for which we are providing
consultation

Organization dynamics

A strategic focus related to the work

Facilitation skills

Being able to work with others to take strategic concepts and
convert them into action and achieved results

Good listening and communication skills

Being able to objectively frame ideas of others into a work plan

Objectivity in handling information gathering and the
associated analysis

A very important point here is that all three elements need to be present for one to be an internal consultant in an effective and efficient manner. The overlap of all three elements is graphically portrayed in Figure 2-1.

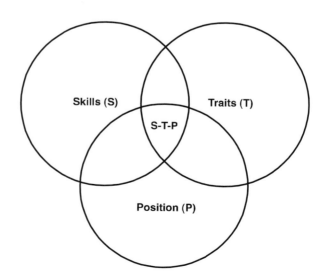

Figure 2-1 The
Three Elements of
Internal
Consulting

Table 2-1

*What if One of
the Elements of
Internal Consulting
is Missing?*

Elements Present	Element Missing	Issues
Skills and Traits	Position	The internal consultants are not properly positioned in the organization. An assignment to a staff level position in a strategic role could solve this problem. Success would also depend on the individual having sufficient credibility to be assigned to this position
Position and Traits	Skills	The internal consultants and their manager need to 1) review the skills needed for effectiveness and then 2) provide the means for the individuals to obtain the needed training or skill development.
Position and Skills	Traits	Coaching to help the internal consultants improve these traits can be effective. Short of this approach, if the individuals do not have the correct traits for the job, alternative courses of action should be considered.

But what if all of the three elements are not present, what are the exposures to effective internal consulting? In Table 2-1, I have addressed the associated problems with one of the three elements missing. If two of the three elements are missing, you need to consider an alternative course of consulting action such as obtaining an external consultant who has both the traits (T) and the skills (S) and provide them with internal support having the positional (P) requirements.

2.5 The Internal Consultant RACI

Internal consultants work at different levels throughout the course of a work initiative. In some cases, they have sole accountability for the task at hand. In other cases, they may not have the accountability, but they have the responsibility to get the actual work accomplished. Still at other times, the work effort may be the responsibility of the site personnel, with the internal consultants in either a consulting role or one in which all they need is to be informed about the progress of the effort.

In order to provide a clearer understanding of the somewhat shifting role of the internal consultant, we need to consider the various internal consulting tasks in the form of a RACI Chart. This tool enables you to look at the tasks associated with a work initiative and the job roles associated with the task's performance. Then using a series of letter identifiers, we can portray the entire set of tasks in a tabular format, as shown in Table 2-2. These identifiers are R for responsible for task execution, A for sole accountability that the task is completed, C for consulting about the task, and I for those who need to be kept informed, hence the RACI chart.

As you can see from Table 2-2, the role of internal consultants changes as the initiative develops from one of high involvement and responsibility at the outset to a more supportive role as the initiative unfolds. For example, the task associated with identifying needs and benefits had the internal consultants working in a consultative role (C) in support of the site management, who has the sole accountability (A) to conduct the analysis as well as the responsibility (R) to complete it.

Later on in the process after the assessment is completed, the

Table 2-2 The Internal Consultant RACI Chart

Task	Senior Mgt	Middle Mgt	Front Line	Consultant
Identify the need / benefit	A, R	R	C	C
Gather information	I	C	C	A, R
Review and analyze data				A, R
Develop recommendations		C	C	A, R
Present results	I	R	I	A, R
Develop next step	A, R	R		I
Develop gap closure work plan	C	A, R	C	R
Execute the work plan	A	R	R	R, C
Train as required	A	R	R	C
Audit the process for sustainability	A	R	I	I
Take corrective action based on audit	A	R	R	I

internal consultants are accountable (A) and responsible (R) to develop, based on their experience, a set of recommendations. However these recommendations are not developed in a vacuum. The site management is consulted (C) to make certain that the information obtained in the assessment is clearly understood and part of the recommended corrective action.

Still later, after the changes have been implemented, senior management is accountable (A) for the corrective action activity with the other levels of management responsible (R) to implement them. At this stage, the internal consultants are only informed (I) about the work so that they know the process is functioning. Site ownership is clearly shown if the process functions in this manner.

2.6 The Internal Consultant Continuum

As the RACI Chart points out, the role of the internal consultant changes as the effort develops. The problem faced by many internal consultants is that they fail to change when the time is required. Many take on the role of the work executioner throughout the effort, leaving as helpers the site personnel who should be building ownership. It is clear that working in this mode will disable the site personnel; in the end, it very likely causes the demise of the change initiative. At some point in time, the internal consultants need to become supporters of the site personnel, who really are the ones who should be doing the work.

On the other extreme are the internal consultants who believe that they are there only to advise and never to get involved with the actual work. This approach will also end in failure of the initiative because the site personnel need guidance at the beginning of the initiative. That guidance is the very reason that the internal consultants are part of the project. Without the necessary guidance, the site personnel may very well head off in the wrong direction, making success of the initiative even more difficult.

The problem is that the internal consultants' efforts are not as clearly defined as one may wish them to be. It takes a great deal of experience to know when to jump in and do the work and when not to. The correct level of involvement, while not being specific, does fit itself into an internal consultant continuum, as shown in Figure 2-2.

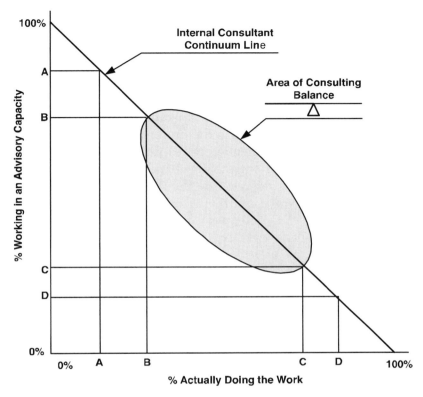

Figure 2-2 The Internal Consultant Continuum

The Internal Consultant Continuum is a chart that shows where internal consultants can be most effective as well as what to avoid. The y-axis shows the level that internal consultants work in an advisory capacity—from 0% at the lower end of the axis to 100% at the upper end. The x-axis shows the percent of actual work done by the internal consultants, again from 0% at the extreme left to 100% at the extreme right end of the axis. The Internal Consultant Continuum is the line connecting the 100% points on the x- and y-axes.

Because the internal consultant's time can only be 100%, they can work at any point along the continuum line. For example, if the internal consultants are working too much in an advisory capacity, the spot on the continuum line would be represented by point A—approximately 90% advisory and 10% working. On the opposite end, if the internal consultants are working too much and not allowing the site personnel the ability to get involved, then they likely would be working at point D—approximately 90% work and 10% advisory. In either case, the internal consultants are at one of the extreme ends of the continuum, a place where they should not be working.

The proper area for internal consultants to be working is between points B and C. This is the area I refer to as the Area of Consulting Balance. It is rather a broad range, but the concept is clearly supported by the various levels of work that the internal consultants perform, as depicted on the RACI Chart. For example, in the assessment phase of the work, the internal consultants would be working closer to point C—doing a great deal of the interviews and analysis work on their own. Later, however, when the recommendations were being implemented, they would be working in an advisory capacity closer to point B. The balance of the varied internal consultant tasks would fall between points B and C, as dictated by the requirements associated with the internal consultants' role at the point in time in the process.

2.7 Internal Consulting Risk

The job of internal consultants is loaded with risk as well as frustration. That is not to say that successfully completing an initiative is not also coupled with rewards and a great feeling of satisfaction, but the risks and frustrations need to be recognized and addressed if one is to succeed.

Risk runs at several levels, as shown in Figure 2-3. At the lowest level is the failure of the initiative on which you are working. If the initiative is an isolated instance and you can move onto the next, all is well and good. Of course, you need to examine why the initiative failed and take the necessary corrective action in order to avoid failure in the future. If the problem is with the work you have done as an internal consultant, then you need to take personal responsibility to improve. If the failure is with the client, then clearly identifying the reason may provide the next initiative with a greater chance of success. Nevertheless, it should be recognized that an isolated failure can be corrected with the next success.

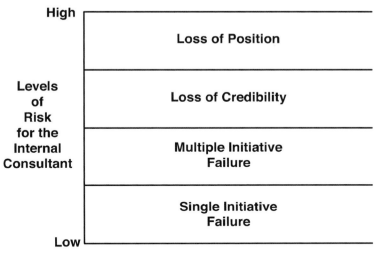

Figure 2-3 The Levels of Risk

At the next level are multiple initiative failures. This has the potential for becoming a more serious issue for both internal consultants and the site. Multiple initiative failures have more far-reaching causes; they are either the result of poor consulting work, site resistance, or both. When there are multiple initiative failures, it is incumbent on both the internal consultants and the site to stop and determine the root cause of the problem. Failure to address the root cause will certainly result in continued failure and, ultimately, both skepticism and further resistance for additional work in the area of change.

The next level of risk beyond multiple initiative failure is loss of credibility on the part of the internal consultants. Loss of credibility means that it is highly likely that you will no longer be used in an internal consultant role. There are two problems here. First is that the internal consultants will no longer be able to continue practicing their craft and adding value to the company for which they work. The second is that once you lose credibility it is extremely hard to ever get it back. Loss of credibility is like a dark cloud hanging over your head. Once it is in place, it is hard if not impossible to remove. To avoid this ever happening, single and multiple initiative failures need careful analysis and corrective action before you ever reach this level.

At the highest level of risk is loss of position. In this scenario, the internal consultants are fired. However, equally bad is to be moved to a position outside of internal consulting which you may not want to perform. In the latter of the two situations, leaving the company is usually the end result due to the high level of frustration that accompanies such a severe action on the part of the management.

2.8 Internal Consulting Frustration

In addition to risk, many internal consultants are faced with frustration as they try to deliver value to their company. If you have offered up a good idea and had it rejected only to have it acted upon when the same idea is suggested by an external consultant, you will understand internal consultant frustration. There are many reasons that this happens, even to individuals who have been the designated role of internal consultant within their firm. These include:

- Lack of a champion in senior management who will support your initiative
- Failure of the site management to recognize the ability and value that you bring with your work knowledge and experience
- The belief that, if the idea is not suggested from outside the company, it can't really be that worthwhile
- Lack of credibility on the part of the internal consultant

For each of these reasons, the suggestions by the internal consultants are not acted upon by the organization, even though they may have significant value. Overcoming each one of these and others that may be causing frustration needs to be approached in a different manner and within the cultural context of the company in which you are working. For example, if you need a champion in order for your idea to be used, then you need to figure out how to get one. Failure to do so in an organization that requires this level of support for new initiatives is imperative. For the other bulleted items, you need to figure out how to 1) get the site management to recognize your ability, 2) believe that internal ideas also can be very valuable, and 3) promote yourself into a position of credibility.

2.9 What Does Success Look Like?

Success comes in many shapes and sizes. Ultimately what success looks like for an internal consultant is determined by four factors:

- Whether or not the internal consultants were able to add value to the reliability / maintenance effort and ultimately the work site.
- Whether or not the initiative is able to sustain itself over the long term with no internal consultant support.
- Regardless of the time and effort expended by the internal consultants, whether or not the site personnel own the result and believe that they accomplished the task on their own with a little consulting support.
- Whether or not the internal consultants live to work another day and are enabled to add continued value to the company.

All of these, when taken together, clearly describe the role of the internal consultant.

Five Things to Think About or Do

1. Identify who in your organization fills the role of internal consultant, either full or part time.
2. If the individuals identified in #1 are internal consultants, what skills, traits, and organizational position do they have that makes them effective?
3. Examine the skills, traits, and job position in detail. Can you see how these things help make them successful in their job? If pieces are missing, can you see how this gap can negatively impact on their ability?
4. Examine the RACI chart in the chapter. Can you see how the various tasks fit with the individual's roles in the

process? If you are acting as an internal consultant, can you see how your role fits with the various tasks listed?

5. What things can you do to mitigate internal consultant frustration or failure? List them and determine how they can become part of the work you do.

STRATEGIC AND TACTICAL WORK

Strategic and tactical work are like oil and water
They both exist but you can't mix them

3.1 You Can't Do Both

Imagine that you are a maintenance superintendent in a plant. Your job entails supervising several maintenance foremen and their work crews as they work to maintain the production lines. Further suppose that the organization of which you are a part is highly reactive in nature. In other words, when things break down, production expects Maintenance to drop whatever they are doing and response to the emergency of the moment. Working in this manner, there is little or no room for reliability practices or programs. As a result of the reactive nature of the business, none exist. Then one day you are summoned to meet with your boss and the plant's senior managers. It seems they have finally recognized that the equipment is very unreliable and that the reactive mode of operation is actually hurting the business.

As a result of this revelation, it has been decided that you should set up an equipment preventative maintenance program to reduce the level of plant reactivity and, at the same time, to improve the reliability of the equipment. To you, this is great news. You have been a reliability advocate for years, but never were able to get the management team to listen to your ideas.

So when you return to your office you sit down and begin to think about how you would set up the plant's preventive maintenance program. Less than one minute into your planning, the phone rings. There is a problem with the main charge pump that requires your attention. You leave your office, handle the problem, and return after a few hours to the design of the program. Soon there is a knock at your door. It is one of your foremen with a personnel problem. It seems that there has been some faulty workmanship and the foreman does not know how to deal with the person responsible. Once again you leave your office, help the foreman resolve the problem, and return several hours later to resume the work you were doing on the preventive maintenance program design. The phone rings. This time, it is your boss, who has to leave the plant for a doctor's appointment and needs you to fill in at a meeting that begins in one half hour. Again out of your office you go to attend the meeting. You return several hours later only to realize two things. First, due to the constant interruptions throughout the day, you have accomplished nothing in the design of the preventive maintenance program. Second, it is time to go home. The next day you begin the process all over again. Very quickly, the entire week is gone and still nothing has been developed related to the new program.

Years ago I held the position of maintenance superintendent in a very reactive maintenance organization. My day, week, month, and year were exactly as described in the above scenario. I came into work in the morning and the first thing I did was stop at the shift supervisor's office to find out what equipment had broken down during the night shift. Then I and my organization spent the rest of the day fixing the failed equipment only to repeat the same process the next day.

Then something amazing happened. I was transferred to corporate headquarters to lead a computerized maintenance management systems project. This project was still in the developmental stages so there was a lot of planning and work redesign required. About a month into the work effort, I recognized that I was working in a very different manner; I spent some time trying to understand why I felt this way.

If you ever have had a time when you were trying to figure out something and the proverbial light bulb went on, that is what happened to me. I recognized that the the work that I had been doing in the plant and the work I was doing at corporate headquarters was different. In my former job, all I was able to do was to react to plant problems. In the corporate job, however, I was able to think and plan for future events to achieve a long-term goal. These two types of work were very different, which explained why I felt the way I did.

This revelation has a direct connection to the scenario I presented at the beginning of this chapter. The superintendent in our example was always working in a reactive (tactical) mode. Consequently, he never had the time to think (strategic) about how he would create the new preventive maintenance program. This is significant; you can do one or the other well but you can not do both well at the same time. This fact is important for all maintenance professionals who struggle to create proactive strategic solutions while locked into a demanding day-to-day tactical work process. It is even more significant for internal consultants.

3.2 Strategic and Tactical Thinking

The two types of thinking that are required in the reliability and maintenance arena fall into the categories of strategic and tactical.

Strategic Thinking

Strategic thinking is more conceptual and long term. The purpose of this type of thinking is to develop plans for achieving the vision and goals set by the organization. It is essentially a series of thought processes designed to figure out how to get from where we are to where we want to be in the future. It can also have details associated with the steps in the process, but they are not often fine tuned; they are more of a general guideline showing how to proceed. Suppose our goal was to develop a preventive maintenance program. Then the strategic process would identify several initiatives as well as a set of strategic-level questions that would need to be addressed if we wished to succeed. Examples of these initiatives and the strategic questions that need to be answered include:

- Develop the scope of the effort
  - What type of PM work will be done?
  - What will be the frequency of the PM effort?
  - How will the work be scheduled and compliance tracked?

- What records will be maintained to show the work was completed?
- Determine who will do the work
  - Will the work be handled by Maintenance or Production, or will it be contracted to a third party?
  - If internally developed, how will the crew be selected?
  - If external, who will identify the contractor and write the contract?
- Determine the required training and tools
  - What skills are required?
  - How will people be trained?
  - What type of comprehension testing is required?
  - Is there a need for a periodic refresher training program?
  - How will new technology be integrated into the program?
  - What tools are needed and how will they be obtained?
- Develop a communication plan for the site
  - How will the site personnel be made aware of the new PM process?
  - What do they need to know regarding their accountability for success?

As you can see, each of these initiatives is rather broad with many strategic questions that must be asked and answered. This allows a great deal of latitude as the site figures out how to make the effort a success.

Tactical Thinking

The other half of the equation, tactical thinking, addresses the step-by-step efforts required to get the work accomplished. In my books *Successfully Managing Change in Organizations: A Users Guide and Improving Maintenance and Reliability Through Cultural Change,* I describe a tool called the Goal Achievement Model. This model addresses initiatives (strategic activities) and shows how each of these needs to be acted upon at the specific activity level (tactics).

The strategic examples shown above also have tactical components in order that they can be completed and drive the work process forward. For example, in the initiative related to the development of the work scope, the following tactical activities may be present:
- What type of PM work will be done?
    - Examine each of the equipment types within the plant and determine the specific PM work that will be done on each one.
- What will be the frequency of the PM effort?
    - Review the manufacturer's recommended PM schedule as well as the plant engineer's experience. Then develop a specific schedule by date for each type of equipment.
- How will the PM compliance be tracked?
    - Develop specific measures to track PM compliance so that any process slips from the schedule will be recognized and corrective action taken.

As you can see, tactical thinking goes hand-in-hand with and supports strategic thinking. In our example, the tactical part of the effort is in the development of specific tasks to accomplish and support the strategic element—the initiatives.

The scenario I presented and my comments about not being able to do strategic and tactical work at the same time may appear to be in conflict with this example, but they are not. Strategic and tactical work can not be done at the same time. That does not mean that they are not linked. But how does this apply to our superintendent in the example: Certainly he is not able to do the strategic work of developing the PM program as he responds to the tactical day-to-day needs of the plant.

There are two parts to this answer. First is that strategic thinking and the resultant strategy creates the need for tactics and activities, but they need to be aligned. In the superintendent example, they were not. The plant's strategy was reactive repair. The superintendent's role was to carry out the tactics to support it, not to develop and implement a PM program. Second, strategic and tactical work can not be carried on at the same time due to the very nature of the work and the vastly different thinking and parts of the brain required for each of the efforts.

3.3 Left Brain—Right Brain

One of the primary reasons that strategic and tactical thinking can not be carried on at the same time is that each is conducted in a different hemisphere of the brain. Experimentation with the various ways human being think has shown that the two hemispheres of the brain—the left and the right—are each responsible for different manners of thinking about things. The left brain is logical, analytical, and objective; it looks at the parts of an effort. These tasks are what you expect from people with the ability to be good tactical thinkers. They take the problem at hand, examine it logically, analyze the results, and take action to correct the parts that are in need of repair. Think about our superintendent and how he was called upon to handle tactical activities throughout the day.

The right brain is more intuitive and subjective; it tends to look at the big picture. These are the traits that one would expect associated with strategic work efforts. Although the superintendent may have been able to do strategic thinking and develop the PM program, the tactical needs of the daily operation precluded his ability to switch sides of his brain, shut down the tactical side, and engage the strategic side.

The real problem is that both sides can not work at the same time. This also explains why, when I was assigned to the computer project in corporate headquarters, I felt different. It was not that I was simply able to think as I initially felt when I examined my situation. It was that I had been removed from my tactical role and not subjected to tactical interruptions. I was then able to engage the right side of my brain and think strategically.

Figure 3-1 depicts strategic and tactical thinking in the form of a quad diagram. The diagram shows the various areas in which people operate. I have marked an X in Quad 1 because we need not consider people who have neither tactical nor strategic ability; they probably won't be around too long. The x-axis measures strategic skills as either low or high; the y-axis does the same for tactical skills.

Those who work in Quad 2 have high strategic and low tacti-

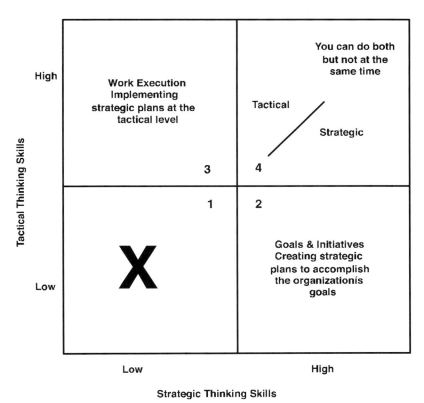

Figure 3-1 The Strategic – Tactical Quad Diagram

cal skills. These are the people you need to work on strategic initiatives that improve the reliability of the equipment. Those in Quad 3 do not possess strong strategic skills. However, once the strategic plans have been formulated, these are the individuals who can make it happen in the plant. They are the ones to get the job done.

In Quad 4 are those individuals who are skilled in both areas. They can think and formulate strategic plans and, if given the opportunity, they can put the plans into action. However, as I have said, these two tasks can not be done at the same time.

As we have discussed, some people can operate in either mode as long as they don't try to do both at the same time. However, what if a company takes a good tactical person and puts them into a strategic role or conversely puts a good strategic person in a tactical role, and they are not suited for the task? In either case, there is immediate frustration on the part of the employee and longer term loss of valuable assets for the company. The way you can identify this problem is as follows:

Tactical people in a strategic role will tend to focus on the details vs the big picture. They will be able to add very little to the development of strategic plans. When asked about the issues they are having with the job, they will tell you that if you tell them what you want done they will deliver it. This inability to contribute to the strategic plan coupled with a real desire to get back on the line and work on day-to-day tasks is highly frustrating; it will lead to long-term job dissatisfaction and loss of valuable company employees.

Strategic people in a tactical job will always be focused on the big picture and the strategy needed to accomplish it. This focus will come at the expense of the day-to-day tactical work that you expect them to be doing. When asked, they will tell you about the really good ideas they have, but you will be more interested in why the daily tasks have not been handled well. Their inability to contribute strategically coupled with their lack of interest in task-related activities will also lead to frustration and the same results as the tactical person in a strategic role.

3.4 The Internal Consultant Dilemma

This brings us to those people who have the potential to think with their left and their right brain. These individuals can not only develop strategic plans, but they can also deliver the results through tactical work in the field. In fact, it is highly likely that during the course of their careers they have had the opportunity to work in both areas. This skill makes them highly valuable to the organization as long as their skills are properly utilized.

In Figure 2-1, we discussed the internal consultant continuum. We showed that there were areas where internal consultants should not be working—either an overabundance of advisory work or actual work in the field (points A and D). We also showed the range where an internal consultant should be working (between points B and C). With this in mind and referencing Figure 3-1, it is clear that the internal consultants should be working in quad 4 and effectively balancing their focus between strategic and tactical work.

Given the fact that you can not do strategic and tactical work at the same time, and given the fact that a good internal consultant needs to work in both areas, how can this be accomplished? Furthermore, what do the internal consultant and the company for whom the internal consultant work need to do to assure a successful outcome?

3.5 The Company

A company, with the best of intentions, can severely hinder the success of its internal consultants. Remember that those who typically fill the internal consultant's role are those who are best at handling what is often referred to as special projects. Then how can a company hinder this effort? The answer lies in the way that the company addresses special projects and those who fill these roles. There are several views of special projects:

- Special or strategic projects are more figments of a company's mind than actually developed and implemented.
- The person in this role has special project abilities but the company does not recognize this as internal consulting nor consider it value-added work.
- The person does not have the skills for reliability and maintenance work in a reactive line organization. Rather than fire the person, the company is using the special project role to move them out of the way.

Many companies state for the record that they are working on special projects in order to help them move away from their current reactive work stance, or to improve upon some selected proactive maintenance strategy already in place. However, this is far from the truth; although the project exists on paper, it does not exist in the real world. In fact, every time that the individual or group assigned to work on the effort is prepared to begin, there is another "emergency of the day" to make it impossible to release them.

The second view of internal consultants is the one where they have been given responsibility to develop initiatives and add value, but they are hindered by the very company that put them in this position. The issue is that the company sees the work as individualized special projects. As a result, the company expect them to have a beginning and ending, at which point additional projects may be assigned. To make matters worse, the company often assign this work to people who already have full-time tactical jobs. This is exactly what happened to the superintendent in our example at the beginning of this chapter. The problem is that the company does not understand what they have really asked the consultants to do. As a result, they set them up for failure.

The company can correct this situation by:
- Recognizing that special projects and the individuals assigned to this work are strategic in nature—and essentially is internal consulting.
- Supporting the individuals assigned to the work by keeping them out of the tactical day-to-day work process. If they tend to get pulled into the day-to-day work, the company needs to help them isolate themselves from this serious distraction.
- Monitoring the work being done and help them stay within the internal consultants' area of balance (between points B and C in Figure 2-1).
- Providing training and guidance, thus enabling the persons in the internal consultant role to grow and add even more value to the business.

The third view often happens to strategically-focused people in very reactive organizations. It also takes place in organizations that want to make a change in personnel, but do not want to fire those who are the current job incumbents.

Very reactive organizations see strategically-focused people as roadblocks to the day-to-day tactical work. Because their focus is on the problem of the day, they do not want to be refocused on how they can do the work better, or even make changes to their current mode of operation to do the work differently.

Strategically-focused people tend to offer these sorts of ideas. As a result, the company assigns them to special projects where they can develop and propose new ideas and still allow the reactive day-to-day work to continue. The problem is that 1) the proposals coming from those in special project roles are often dismissed or put on the back burner, and 2) those occupying these positions often become frustrated and leave.

Organizations that want to make a change in those who currently hold positions of influence often assign these people to meaningless special projects. That enables the company to move them out of the way without the guilt of firing them. In the end, the individual leaves due to frustration and the company has achieved its desired end. I once worked for a person who ran the field execution part of the maintenance organization. He and the maintenance manager did not get along so he was assigned to a series of special projects; his line position was given to someone aligned with the maintenance manager. The projects that he was assigned were never going to be implemented. Everyone including the individual in question knew it. As a result, it wasn't too long before he quit.

3.6 The Internal Consultant

Those working in the role of internal consultant can also hinder their own ability to add value to the work in which they are engaged. There are many ways that this can happen:

Not making sure you get what you need to do the work. Because internal consultants are involved with the work initiative in a supporting role, there are several things that they need to make sure they have so that they don't derail the effort. You could say that these are things that the company should provide and you would be right. Many companies want those within their organization with internal consultant skills to function in this role. However, they don't know how to go about it, or they don't know what needs to be provided to the internal consultants for success. Therefore, it is incumbent upon you, the internal consultant, to make sure you clearly identify what you need and make sure that your company provides it. If not, you have severely hindered your ability to achieve the outcome you seek.

- For example, you need a member of senior management as your sponsor. Without a sponsor, the work you are doing will never get to the correct level of the organization that could take ownership and support implementation. You also need to make certain that you and those that are part of the change team are given the time, money, and resources to get the job done. Remember you are there to support the site. This specifically includes site ownership. Without time, money, and resources, the initiative will never get off of the ground. Those involved, and maybe even you, will be moved to other "more important" efforts.

- The last thing that you need to make sure you acquire is clarity about the work that you are expected to perform and the time allotted to complete it. Without obtaining clarity, you may very well work on the wrong thing and deliver the wrong and unasked for result. Again you would think that clarity should be something provided to you, and often the company, your client, does provide it. However, as with all communications, you need to make sure that you and your client are clear and in agreement regarding exactly what you are expected to work on and deliver. The other aspect that is important for you and the client to agree upon is timing. It is wonderful to deliver a great product but not so wonderful if it was needed last week and you show up a week late.

- Working too much in Quadrant 2. Many people feel comfortable working in this quadrant because the work is oriented toward the future and developmental in nature; they like working in this manner. Quadrant 2 addresses how things will look in the future and creates plans to accomplish the vision.

However, consultants can find themselves working too much in this area and not paying as much attention to the hands-on work as might be needed. This imbalance hinders their ability to be fully effective and, if self imposed, needs to be corrected. Not only does the focus need to be changed, but the internal consultants also need to examine why they have set themselves in this position. Is there a reason that they are avoiding the hands-on part of the process? If so, recognition of the reasons behind these feelings will help them adjust the balance of their efforts. If it is due to the state of the work—more strategic effort is required—then they need to still pay attention to how they balance their efforts and be prepared to shift roles when the time is right.

- Allowing yourself to be dragged into Quadrant 2 when you should be elsewhere. Not only do internal consultants often spend too much time in Quadrant 2 by design, they often get dragged into this role when they really need to be work ing in other parts of the quad diagram. When this occurs, the internal consultants needs to determine why they allowed this to happen, essentially reducing their ability to add value. If they really belong in Quadrant 2 and have been dragged there by the process for a reason, all well and good. If not, the reasoning behind why they allowed this to hap pen will allow for corrective action on the part of the internal consultant.

- Working too much in Quadrant 3. The same rational that applied for working too much in Quadrant 2 (strategic) equally applies to those internal consultants who spent too much time working in Quadrant 3—actually doing hands-on tactical work. Once again, the reasoning needs to be reviewed. If there is a self-imposed imbalance, corrective action on a personal level is needed.

- Being dragged into Quadrant 3 when you should be else where. Allowing yourself to be dragged into Quadrant 3 when you should be working elsewhere is a problem on two levels. First is there a reason why you as the internal consultant have allowed this to happen. Would you rather do the work than develop strategies for achieving the vision? To be an effective internal consultant, you need to balance your activities; if you have allowed yourself to get too deep into the actual work, you need to understand why and work towards creating the personal balance.

- Trying to work in both parts of Quadrant 4 at the same time. The issue here is that, as an internal consultant, you have not recognized that you can not work in both the strategic and tactical roles at the same time. Once again, internal con sultants need to create a sense of balance, understand the needs of the client, and position themselves in either the strategic or tactical roles as the need arises.

3.7 The Balance: Left Brain – Right Brain

Left brain (tactical thinking) and right brain (strategic thinking) are two processes that are mutually exclusive. You can do one or the other, but you can not do them both at the same time. Nevertheless, it is the job of internal consultants to be able to employ both left and right brain thinking in the proper balance to adequately perform their role within their company. Although this certainly is no easy task, it can be accomplished by identifying what role you are trying to fill and then applying the correct side of the brain to the effort. The trick is to be able to do this effectively and efficiently as you work in support of your client. The other side of the coin is also equally important. That is to be able to recognize when the wrong side of the brain is engaged as well as to make the effort to correct the mismatch. Without these capabilities, the real value of the internal consultant is lost; as a result, both the internal consultants and the client suffer.

Five Things to Think About or Do

1. Do you clearly recognize the difference between strategic and tactical work? Provide yourself with some examples of each in your work.
2. Have you ever tried to conduct strategic work while being immersed in the day-to-day tactical world? What happened? Were you able to accomplish both types of work successfully at the same time?
3. Are you a tactical or strategic type thinker? Identify the issues you had (and what you did about them) when asked to work in the opposite role. Think of examples provided by others who were miscast. How did they do working in a role for which they may not have been suited?
4. Identify ways that your company has inhibited the successful completion of strategic projects. What should have been done differently?
5. Identify ways that you (acting in the role of internal consultant) have inhibited the successful completion of strategic projects. What should you have done differently?

The Business Case
for Internal Consulting

*We know that internal consulting is needed
for successful change
The business case, if done correctly,
proves it to the doubters*

4.1 Why Is a Business Case Needed?

Most companies have internal consultants, except they often do not call them by this name. The label often given people who hold these positions is engineering or reliability staff—people who work in other departments outside of maintenance whose role it is to provide support to those doing the work. If the role is filled from within the reliability / maintenance organization, either on a full or part-time basis, these people are often described as simply handling special projects. The projects can be of either short- or long-term duration.

In each case, these people provide the services that we have been describing as those delivered by an internal consultant. The problem is that these individuals do not have the correct label assigned to their role. As a result, the incumbents often find it difficult to work in the role to which they have actually been assigned. What is needed is a way to validate the position of internal consultant, then fill the position with an individual or group whose sole job it is to perform this service for the organization.

Years ago I was working as a lead engineer for a section of the plant. This job was not one that I really wanted; I had landed in this role as the result of a company-wide layoff. My options were lead engineer or seek work with another company. The reason I mention this was that this position was not something that needed my full-time attention due to the fact that the engineering was actually being handled by my staff. My role was one of coordination and it was not fully time consuming.

As a result, I began having discussions with a good friend of mine who at the time was the manager of the Maintenance organization. Over the prior three years, he had become increasingly unhappy with the current maintenance work process. He had decided to redesign the work flow, the organizational structure, and many of the other components that were required for a successful, reliability-focused maintenance effort. With time on my hands, I agreed to help facilitate the effort. In addition to my experience in the maintenance arena, I was also able to deliver something else that was even more important. I did not have any line responsibilities. As a result, I was able to focus my efforts on the redesign initiative. As I was engaged in this initiative—which took over six months to complete—something was happening to me that I had not yet identified. I was becoming an internal consultant for the organization.

After a great deal of effort on the part of many people within the organization, we developed and deployed an entirely new way of conducting maintenance within the plant. At the same time, the plant was also going through a total reorganization of which the maintenance organization was a part. The maintenance manager created a position reporting directly to him identified as Special Projects Coordinator; I was assigned to this newly-created role. In the back of both of our minds, we recognized that having someone who could detach himself from the day-to-day tactical work of the organization and work strategic (special) projects had the potential of providing immense value to the organization. Still we had not identified this as an internal consultant position.

Over the next several years, I worked for the maintenance manager in this role. The problem that I had was that all of my assignments originated with him. While I did offer my services to others, they did not fully understand what I was capable of delivering. Nor did they see the value that I could deliver in a special project role.

For line organizations that are focused on reactive maintenance, which mine was, this failure to recognize value from strategic work initiatives delivered by someone focused in this area is not uncommon. Strategic work is something that they do when they have some spare time, which never seems to happen. Their focus is on the day-to-day; focusing on longer-term initiatives is not what they get rewarded for doing.

After the maintenance manager retired, I worked for several other managers who saw the value in what I was doing. As a result, I was allowed to keep working in my special projects role. Then something very interesting happened. In a move to further reorganize the department, I was asked if I could use some help handling the many initiatives that my managers had assigned me. Thus the organization grew from one to two people. This happened several times over the next few years and slowly I acquired an organization of people working on plant-wide strategic initiatives. At the same time, the organization itself evolved into one that placed a much higher value on reliability. This new direction supported the work my team was doing. Eventually my small organization was recognized as delivering strategic value and it was sought out to support the reliability / maintenance organization as well as others.

The reason that I have related this story is that my journey from part-time special projects coordinator to one where my team acted as reliability/maintenance internal consultants—even though we were not named as such—was due to fortunate timing and a lot of luck.

Placing people with strong internal consultant skills within the reliability/maintenance organization in order to support the development and implementation of strategic initiatives should not be left to chance. There needs to be a better way, a process that identifies the need for such a position and then goes about the task of justifying it. If this is handled correctly, then the individual or group appointed to these positions will have 1) a clear mandate from the site management, 2) recognition by the site personnel of their role within the scheme of plant operations, and 3) identification of the individual or group as internal consultants for the plant. This is clearly a better way to reach internal consultant status than banking success on timing and luck. The question is: How can this be accomplished? The answer is that a clear business case must be developed, submitted, and approved by site management, thereby sanctioning the role.

4.2 What Is a Business Case?

A business case for an internal consulting role, or for anything that you wish to have sanctioned by management, is a written document that clearly delineates the need for what you are seeking and provides an explanation of the value that can be attained by getting agreement to proceed. Suppose that you have a pump that is continually failing, causing interruptions to the manufacturing process and loss of profit. In this instance, your business case would explain the loss to the company with specific supporting data. It also would provide a detailed estimate of the cost of replacement, showing that it was less expensive over the long term to buy a new pump than to keep repairing the old one.

Another example of the reasoning behind a business case lies with a group reorganization. For our example, let's consider reorganizing from a decentralized maintenance organization to one that is centralized—all of the mechanics are dispatched from a central scheduling location rather than being assigned out into the production areas. This is a major change for an organization. A business case will serve to explain the reasoning behind the change and the benefits that the change will deliver to the organization. In this case, the work crews would be assigned to plant-wide priorities vs. those of the individual production areas, which may not be high priority across the plant.

The business case for internal consulting is far more difficult than getting permission to buy a new pump or to reorganize the department. The reason for this is that most of what an internal consulting function delivers is either intangible or is a process that enables other process changes that, in turn, can deliver the bottom-line value that management usually seeks. However, if you wish to have the internal consulting role sanctioned and supported, a business case is absolutely required.

Suppose the outcome of an internal consulting initiative was improved communications between Production and Maintenance. In this example, the poor communications between the two groups was hampering reliability of the equipment as well as the effective-

ness and efficiency of both organizations with relation to maintenance repair. Through the consulting initiative, the root-cause reasons for the poor communication were discovered and corrected,
leading to improvements in both reliability and maintenance effectiveness.

The value attained through improved communication between Maintenance and Operations can be measured and built into
the business case for the initiative. However, how do you measure
and quantify the value delivered by having an internal consultant
support the work effort? Everyone recognizes the value of internal
consulting, yet it remains an intangible. In fact, some would say
that value was not added at all and the improvement would have
been achieved with or without the internal consultant's presence.
Therefore, an internal consultant business case is required; otherwise, the role that the internal consultant plays could easily be
overlooked or minimized.

4.3 The Components

A sound business case needs to addresses itself to six key elements. Some business cases address all of these elements, others
just a few. How many you address depends on what you are trying
to justify. The six elements of a business case are:

- Resource reduction
- Resource re-deployment
- Reduction in external spending
- Revenue enhancement
- Other efficiency and effectiveness gains
- Intangibles

Let's briefly discuss how you would use the six element model
to justify an internal consulting role within your company.

Resource Reduction

Many companies get into a position where they have to go
through resource reductions. Lost contracts may reduce the need
for the current workforce. The company may face required cut-
backs, or even fighting for the very survival of the plant. Working

on a resource reduction initiative can be extremely difficult for someone who is an integral part of the plant workforce. I

Internal consultants can be used because they understand the big picture, can be somewhat objective, and are in tune with the needs of the plant personnel. They can have a positive effect on the work processes and what needs to be done to keep the work flowing after the change. They can also work with the organization to help minimize the overall impact on the process and the people. Those leading the organization are too busy orchestrating the change, and external consultants are viewed as outsiders. Although resource reduction occurs infrequently, internal consultants can clearly help minimize the impact.

Resource Re-deployment

Redeployment of resources is much more likely than a force reduction. In fact, the majority of work process changes involve some sort of redeployment of the existing resources. Because the internal consultants are very involved with process changes, they serve a critical role in resource redeployment as well. Once again, the site leadership is too involved in the actual day-to-day tasks associated with the change. It is therefore up to the internal consultants to work closely with the site personnel to assure that the resources are correctly deployed to support the change initiative.

Several years ago, I was involved in a work process redesign which moved the organization from decentralized to centralized planning and scheduling. As part of this effort, there also was a need to integrate the maintenance crews who, to this point, had spent years in their own work areas. These crews seldom if ever communicated with one another. The original plan was just to bring the crews together into the centralized area, but still dispatch them without changing the crew mix. As an internal consultant, I explained the longer-term value of mixing the crews, thereby allowing the total knowledge of the group to expand. Had I not added this value, a significant opportunity would have been lost to the organization as they redeployed.

Reduction in External Spending

Internal consultants can also add value by showing how improved effectiveness and efficiency of the in-house workforce or in-house work processes can reduce external spending on contractor support. As work process initiatives are put into place, those immediately involved often fail to see that there are new and different ways to perform the work. They also fail to see that, if these new techniques were put into place, the spending on contractors and other third-party organizations can be reduced or even eliminated. This is an area where internal consultants working with the organization can help them adopt new techniques and reduce their external spending.

Suppose the maintenance organization has 100 people performing craft work within the plant at an effectiveness rate of 40%. Also suppose that due to the workload the same plant has a contractor workforce staffed to augment the in-house organization. If you, as an internal consultant working with the site personnel, are able to improve effectiveness by 10%, you will help the plant raise productivity by 80 hours per day, 400 hours per week, or approximately 2000 hours per year. With this increased productivity, you would be able to reduce the contractor workforce by 250 mandays per year—a significant reduction in external spending.

Revenue Enhancement –

Internal consultants can help improve the revenue of a plant. For example, take any of the reliability initiatives in which an internal consultant can become involved. When successfully delivered, each of these will increase the reliability of the equipment, thereby enabling the plant to meet the demands of the marketplace better. Think about plants that operate 24 hours per day, seven days per week. These plants typically do not have the ability to make up for lost production as a result of equipment failure. Hence, unplanned shutdowns cost money. By supporting improved reliability initiatives, internal consultants support a site's ability to enhance its revenue stream.

Other Efficiency and Effectiveness Gains

The work of internal consultants has the ability to deliver intangible benefits centered in the area of soft skills, also known as the eight elements of change. These elements—leadership, work process, structure, group learning, technology, communications, interrelationships, and rewards—are all key components of internal consulting initiatives. Yet they are almost impossible to quantify as you can with the first four components of the internal consultant business case. Nevertheless, they are of critical importance. Positive changes in any or all of them will deliver immense benefit to the organization. These are the elements we will discuss in the next chapter. They also have been discussed in my other two books: *Successfully Managing Change in Organizations: A Users Guide* and *Improving Maintenance & Reliability Through Culture Change.*

Intangibles

As you will learn in Chapter 5, there is another component of the change process that, along with soft skills, is often overlooked in the development and deployment of change initiatives. This component is the organization's culture, made up of four elements—organizational values, role models, rites and rituals, and the organization's infrastructure. The organization's culture is an intangible component of every change initiative. As an internal consultant, you can make sure that the culture and the related impact on the culture from any change initiative are addressed and not allowed to undermine the effort.

4.4 What an Internal Consultant Business Case Looks Like

The business case is a written document making the argument for instituting the role of internal consultant within the Reliability / Maintenance organization. That role is specifically designed to work with the plant personnel to deliver improvements in the plant's reliability and maintenance work processes. In some cases, the business case is trying to identify the need for a new position where even a special projects role did not exist in the past. In other cases, the business case is attempting to have the special projects role solidified and recognized for what it actually is—internal consulting.

A business case needs to include all of the information necessary to obtain management's agreement to put this position permanently in place. Without this identified position, the plant will continue to borrow people from their day-to-day work on a part-time basis or, even worse, have them work on special projects "in addition to the regular job." We learned in Chapter 3 that these approaches are seldom if ever successful. That is why a permanent position is required.

The internal consultant business case has the following sections:

Purpose

This part of the document introduces the issue and states what the balance of the document will be trying to accomplish.

Background

This section identifies the current situation. Often it is used to point out how part-time work in this area has failed to achieve the results that were expected. Specific examples can be used to support this information. The background sets the stage by identifying the need for the internal consultant role.

Benefits

In this section, you state the benefits that can be achieved by having this role as part of the Reliability / Maintenance organization. You need to state both the short-term and long-term benefits. The short term benefits, which may include a dedicated person supporting a current work initiative, are usually easy for management to recognize because they deal with concrete work efforts that are in progress. The longer-term ones which do not address concrete initiatives may be harder to articulate. However, they have far-reaching impacts, not just for yet-to-be-identified initiatives, but also for sustainability of those current in progress. For the benefits section, you should focus on the six components identified in section 4.3. Another approach is to explain the value lost by not having this position. But this last approach is far less likely to succeed than showing the value the job will deliver.

The Process

In this section, you explain how the internal consultant job will be put into place. If someone is already doing this work full time, the official movement to the internal consultant role will be easy. If not, you will need to outline a strategy for creating and filling this position. This is a much more difficult task. Another aspect of the process is the reporting relationship for the internal consultant. Because this role is essentially designed to be an advisor to management, it makes sense that the consultants will report at the highest level within the organization possible—in this case the Reliability / Maintenance manager. This relationship will also free the internal consultants from working at a lower level in the organization and not having their ideas reach the manager.

The Cost

There is always a cost associated with establishing a new position. The justification for this cost must be tied to the benefits section so that a true cost–benefit analysis can be performed. This justification is not difficult if the internal consultant is dealing with initiatives that have real documented savings. It is, however, very difficult if the work is in the area of intangibles and neither direct cost savings nor avoidance are apparent.

What Success Looks Like

This section is designed to close out the case. It should paint a clear picture for the managers so that they can visualize what having an internal consultant on their payroll looks like. Tie this section to the cost and benefits sections to establish the link.

4.5 You Need a Business Case for All That You Do

Let us assume that you have achieved your goal and gotten approval to establish a position for an internal consultant. Even with the internal consultant business case accepted and the role of internal consultant approved, there are many other instances where the business case approach is needed in order to have internal consulting included within major work initiatives.

As an internal consultant, you need to make sure that you are involved in support of new change initiatives because of the value you can deliver. These initiatives are often presented and accepted by management in the form of a scope of work (see Chapter 8). A typical scope for a change initiative usually addresses the work to be done, how it will be done, and the expected benefits. Seldom is the issue of internal consultant support included. In fact, it is more often completely overlooked. This area is where the scope needs to be expanded and a business case for an internal consultant resource included. The same outline we used for the justification of the internal consultant can be used to assure that an internal consultant is an integral part of the change initiative.

A business case within the work scope also needs to be made for the identification of a change initiative champion. It has been shown in numerous cases that when an initiative has a champion from the plant's senior management team, the likelihood of success is far greater then when one does not exist. This follows from the concept that "what gets measured gets done." In this case, it is not so much that the work is getting measured, but rather that it is getting the level of attention and support required for success. As we will learn in Chapter 5, the champion essentially is tying the initiative into the values of the organization's culture.

4.6 Presentation and Buy-In

The presentation of the business case is critical, whether it is for the establishment of an internal consultant position, having an internal consultant made a key part of the change team or obtaining an initiative champion. You usually get only one chance to make your case, after which the decision is made and you have to deal with the outcome.

In determining your approach, understand how your organization works and proceed accordingly. I have worked for organizations that never made decisions at meetings, even though on the surface it appeared that they did. In reality, you needed to make your case with each manager separately. The meeting was then just a final review and approval. Before I learned this, I made a presentation requiring a decision at the meeting and was unsuccessful in gaining acceptance of a very important project.

For other companies, making decisions at meetings is how business is conducted. This environment requires a vastly different approach in order to achieve a successful outcome. The important thing is to assess how decisions are made within your company, then present your business case with the understanding of how things work.

Often as you go though his process, those from who you seek approval make suggestions that will require you to do additional work. For instance, you make a presentation requesting approval to develop an improved work planning process which, according to your business case, will clearly add immediate value to the business. At the conclusion, one of the managers asks if you have taken the time to see how the competition is handling this process.

On the surface, this appears to be a logical question. But is it? It could also be what I refer to as a deflection, an additional task that 1) is not easy to accomplish, and 2) will make you and your request disappear, at least for some period of time. The deflection needs to be addressed, if possible, so as not to lose time and momentum of your effort. Asking questions to get clarity concerning the request is often a way to accomplish this task. Ask the requesting manager why they feel this search will add value. Ask them what they expect you to accomplish by taking the time needed. The answers to these and other similar questions may eliminate the need for you to stop what you are doing while you go off in a hunt for additional, but meaningless information.

4.7 Showing Value After Acceptance of the Business Case

Managers who accept the business case that establishes the internal consultant position expect return on their investment. How do you provide this return? The answer is to utilize your internal consultant skills to assure that all your initiatives are handled professionally, have a high degree of quality built into the design, are successfully deployed, are sustainable into the future, and deliver the value for which they are designed.

Five Things to Think About or Do

1. If the position of internal consultant does not exist in your organization, yet you are filling this role, consider developing and presenting a business case to management to get them 1) to recognize the position and 2) to assign you as the site internal consultant.

2. Consider the six elements of a business case; resource reduction, resource re-deployment, reduction in external spending, revenue enhancement, other efficiency and effectiveness gains and intangibles. Identify specific initiatives in which you have been involved that delivered value in these six areas.

3. Identify how your organization makes decisions. Do they work as a team and make decisions at team meetings? Do they need to have a one-on-one discussion prior to the meeting so that the meeting is simply a rubber stamp? Is there a different approach that they follow? Knowing this is important to the internal consultant.

4. Have you ever developed a business case document to use as a tool to get management approval of a strategic initiative? If yes, were you successful? If not, why?

5. Find examples of successful business cases within your company. Use them as templates for when you need to develop a business case for an initiative.

SOFT SKILLS AND THE ORGANIZATIONAL CULTURE

A change process that only addresses the hard skills
Is going to be a hard if not impossible process to
successfully implement

5.1 Change Process Failure

Have you ever worked on an initiative that you and others involved firmly believed would deliver significant value to your organization? The scope, the process for deployment, and the commitment from the site leadership were in place. Other companies had implemented the same change and there was evidence that their outcomes were significant in terms of cost savings and reliability improvements. The problem that you were addressing was clearly defined and the solution you were delivering was specifically designed as the solution. However, several months later, implementation problems appeared which became difficult to resolve, even with a great deal of work from the project team and others. A year after the implementation, your initiative not only had failed, but it was also virtually impossible for anyone to find evidence that the initiative had ever existed.

Most readers have had this experience. The question is: What happened to make your effort virtually disappear from your organization's landscape? For the management team and others who were counting on the benefits and may even have presented the potential improvement to senior management, the implications of failure can range from frustration all of the way up to a career-ending experience. For you as an internal consultant, it can harm your credibility or, even worse, your career. That is why it is so important to understand what takes place in initiative failures so that these issues can be addressed proactively, reducing or eliminating the failure.

Many years ago I was involved in a major project to implement a proactive maintenance program involving weekly scheduling, detailed planning of the work, and work crews dedicated to performing according to the plan with very tight controls on deviations. The team that developed the work processes, the training program, and the deployment process spent a great deal of time and effort in these tasks. They had senior management support, a dedicated team of people who were the users of the current reactive work process, and both internal and external consultant support.

With advanced communications, the process was deployed. Those involved followed it as they were trained. However, within a few weeks problems began to appear. Some of the problems included; 1) work crews were diverted to emergency jobs, even though many of these jobs turned out not to be emergencies at all; 2) the planners were not able to plan because the foremen continued to use them on day-to-day task support; and 3) Production disrupted the schedule for jobs that were of "higher" value—even though these jobs were in the backlog and not put on the schedule when it was developed.

As these problems began to materialize, the members of management who had sponsored the event were focused on new initiatives; they did not provide the support required to address the issues in a timely manner. The users, who had been part of the work team, had returned to their regular jobs and were not available any longer. The external consultant had left because the project was finished, and the internal consultant had no authority to address the problems, let alone to take any corrective action. The result of all of this was what we would expect. Six months after deployment, the site was back to its reactive work processes, the planners were providing field support—not planning, and the scheduling process was a joke.

How did a project that was recognized as an important and value-added effort for the organization fall apart like this one? The answer to this question is critical for everyone to understand, especially someone working as an internal consultant. Failure to understand and address the root cause so that corrective action may be taken will cause the initiative to eventually wind up in the corner gathering dust.

The main reason for failure of maintenance and reliability initiatives is that efforts of this type only address the hard skills of the organization, but fail completely to address the soft skills and the organization's culture.

5.2 Soft Skills

In order for change initiatives to be successful, it is critical that all aspects of the process be integral parts of the design. This typically doesn't happen in the majority of change efforts that take place. What does happen is that the organization develops and works on their change initiatives at what is called the hard skill level. Hard skills are the task-oriented efforts that we execute as part of our day-to-day jobs. They include areas such as planning, scheduling, and work execution.

Figure 5-1

Unfortunately, the result of working only at the hard skill level is most often failure of the change initiative. This is a two-fold problem. First, the effort into which the organization has poured so much time and effort fails to achieve the goals and benefits that were established at the outset. Second, and even worse, the failure leaves a sense of skepticism within the organization, making future change efforts even more difficult. There is another level below hard skills that must be recognized and addressed. This foundational level is referred to as the level of soft skills (see Figure 5-1).

The Soft Skill Level

The soft skill level is made up of a set of key elements called the Eight Elements of Change. These include leadership, work process, structure, group learning, technology, communication, interrelationships, and rewards. There is a detailed description of these elements in my book *Improving Maintenance & Reliability Through Culture Change* where each element has a full chapter dedicated to providing the reader with more detailed information.

When you think about the Eight Elements of Change, you can quickly recognize the value that correctly addressing each one can have to your overall change effort, as follows:

Leadership

Leadership is the keystone of the soft skills and of the Eight Elements of Change. Our leaders, whoever they may be, clearly set the tone for all initiatives that are undertaken. If the leadership visibly supports a change initiative, then it is highly likely that money, resources, and the other factors required for successful change will be provided. Conversely if the leadership is not invested in making the change, the likelihood of success will be severely diminished. The leaders provide the direction, guidance, and support that enable the organization to undertake and deliver change.

Work Process

Work process is how our reliability and maintenance work gets performed. There is a drastic difference between reactive- and reliability-focused work processes. Therefore, making a change of this significance is extremely difficult for an organization to undertake. Changing work process not only brings into play working differently, but also has a major impact on all of the other elements of change as well.

Structure

Structure is the organizational hierarchy, including the linkages that reflect how the various units interact. As with work process, there are structures that clearly support reactive maintenance and others that support an organization that is reliability focused. They are vastly different. Changing the structure of a reactive organization, in which everyone understood their roles and responsibilities, to one that is proactive will be a difficult task. However, it must be addressed as part of the change process if the process is to be successful.

Group Learning

Group learning is the ability of an organization to learn from their efforts and then to adjust. This is harder than it appears because we are not asking the organization to simply adjust their activities to be in line with their goals. In addition, we are asking them to re-examine and, as needed, adjust their basic goals. This can be very challenging, especially when the re-examination requires them to alter their basic organizational values and their approach to the work.

Technology

Technology includes the software applications that are utilized to support the reliability focus of the organization. Change often requires various levels of effort in the area of technology. We must first make sure that the applications we are using support the organization's focus after the change. We must then make certain that the functionality within these applications is optimally used by the organization to accomplish its goals.

Communication

Communication is also a critical element of the change process. We may design an excellent process that has the potential of delivering immense value. However, if our intent is not clearly communicated to the organization, the initiatives will break down. In addition, we must not just simply communicate the information. We must also make sure that what is communicated is clearly understood; otherwise, change initiatives are impossible to carry to a successful completion.

Interrelationships

How people interact and work together is an equally important element of the change process. The majority of the work we do in the arena of reliability and maintenance is not done in isolation. For this reason, we need strong interrelationships among the various work groups and the people in these groups. Without these interrelationships, even the best change initiative will fail.

Rewards

Rewards reinforce worker and work group activities. Although it is often believed that money is the ultimate reward, research has proven that this is often not the case. Job enrichment, group acceptance, as well as praise for a "job well done" are other powerful reinforcements. Therefore, it is important to recognize that rewards are needed to support the change effort, and to make certain that the correct rewards are selected.

It is easy to recognize how each individual element of change plays a critical role in the change process. It is also very important, however, to recognize that the elements are far more important when taken together as a dependent set. This is clearly demonstrated when you consider the effect that a change in one of the elements can have on all of the others.

For example, a change in leadership can and often does impact the organizational structure, the work process, how the organization learns and applies the learning, technology and how it is used, communications, interrelationships between individuals and departments, and ultimately how people are rewarded. In other words a change in leadership affects all of the other elements of change. This dependency equally applies to a change in any of the eight elements and the impact on the others. Figure 5-2 indicates the problem that missing one of the eight elements of change in the design of your change initiative can have on the outcome of the effort.

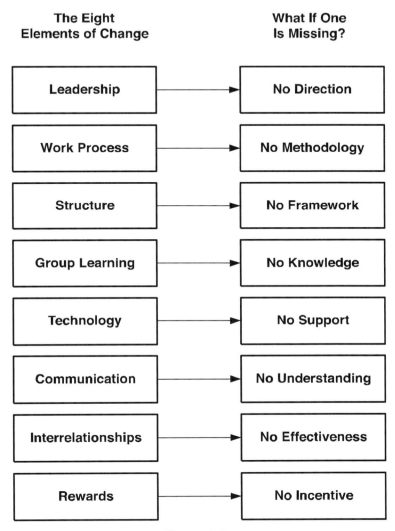

The Eight
Elements of Change

What If One
Is Missing?

Leadership	→	No Direction
Work Process	→	No Methodology
Structure	→	No Framework
Group Learning	→	No Knowledge
Technology	→	No Support
Communication	→	No Understanding
Interrelationships	→	No Effectiveness
Rewards	→	No Incentive

Figure 5-2
What if One of the Elements is Missing?

5.3 The Organizational Culture

For those of us trying to improve reliability or implement any type of change in our business, the question we need to ask ourselves is beyond the eight elements of change. Are there other reasons that can explain initiative failure? The intent of the change that was developed was sound. It was developed with a great deal of detail, time, and often money; the work plan was well executed. Yet in the end, there is nothing to show for all of the work and effort.

Of course, part of the answer is that change is a difficult *process*. Note that I didn't say program, because a program is something with a beginning and an end. A process has a starting point—when you initially conceived the idea—but it has no specific ending and can go on forever.

Yet the difficulty of implementing change isn't the root cause of the problem. You can force change. If you address the eight elements of change in your design, monitor the process, and take proper corrective action, then you may even be able to force the process to be successful over the short term. Here, the operative word is you. Suppose you implement the previously-mentioned preventive maintenance program and then, in order to assure compliance, continually monitor the progress. Further suppose that you are a senior manager; you have the ability to rapidly remove from the process change any roadblocks the change encounters as

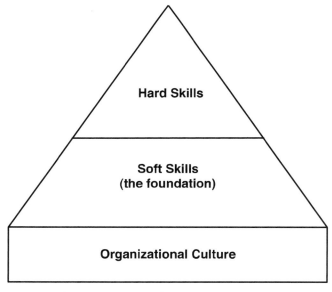

Figure 5-3 The Organizational Culture

it progresses. What then? Most likely the change will remain in place as long as you are providing care and feeding. But what do you think will happen if, after one month into the program, you are removed from the equation? If there are no other supporters, the process will most likely lose energy. Over a relatively short time, everything will likely return to the status quo.

The question we need to answer is why does this happen to well-intentioned, reliability-driven process changes throughout industry? The answer is that the process change is a victim of the organization's culture. This hidden force, which defines how an organization behaves, works behind the scenes to restore the status quo unless specific actions are taken to establish a new status quo for the organization. Without proper attention to organizational culture, long-term successful change is not possible. As shown in Figure 5-3, culture is the third level of the change pyramid, below that of soft skills.

Culture Defined

The starting point of our discussion is to define organizational culture in a way that is understandable. In his book *Organizational Culture and Leadership*, E. Schein defines organizational culture as follows:

> A pattern of shared basic assumptions that the group learned as it solved its problems of external adaptation and internal integration that has worked well enough to be considered valid and, therefore, to be taught to new members as the correct way to perceive, think, and feel in relation to those problems.

If you think about this definition, it clearly describes that subfoundation upon which an organization's behavior is built. It also paints a clear picture of how ingrained these basic assumptions are which, in turn, allows you to understand how difficult they can be to change. Let's look at the component parts.

•"A pattern of shared basic assumptions"

The operative word here is that the culture is constructed upon shared basic assumptions. Because they are shared, when you try to change the assumptions, you need to change them in everyone.

• "The group learned as it solved its problems of external adaptation and internal integration"

The next part of the definition explains that these assumptions are not new creations. They have been tested over time as the organization learned how to solve both the internal and external problems that quite often were serious threats to their very existence.

• "That has worked well enough to be considered valid"

Furthermore, these assumptions worked well for the organization, which has collectively considered them valid. Think about the problems you will face trying to implement change where the new initiative is in conflict with a basic assumption that has been validated over time.

• "Taught to new members as the correct way to perceive, think, and feel in relation to those problems"

This last part locks the assumption into the culture because it is taught to all new members as the "way we work around here if you want to succeed."

This is a very powerful definition if you think about its far-reaching impact on new change initiatives. It essentially says that if 1) a new initiative conflicts with a basic assumption that was learned over time, 2) it has worked well enough to be held as valid, and 3) it is taught to the new members so that everyone believes it as true, then changing things is going to be a very difficult task.

The Elements of Culture

In my book *Improving Maintenance and Reliability Through Cultural Change,* I introduce the four elements of culture—organizational values, role models, rites and rituals, and the cultural infrastructure. These four elements, working in conjunction with one another, make up that rather elusive thing that we refer to as organizational culture. The important thing to recognize is what people really mean when they talk about cultural change. They mean that they wish to 1) alter the value system, 2) displace people who are emulated, but are not in line with the new values, 3) change the rites and rituals, and 4) reframe the cultural infrastructure. Think about the implication of this change effort. It certainly is a major step for any firm to take, which is why it is so difficult to implement and make stick over the long term.

Organizational Values

Organizational values are the beliefs and assumptions that an organization believes to be true; it uses these values as a set of guiding principles for managing its everyday business. The values are what collectively drive decision making within a company.

For instance, an organizational value may be that production is the only thing of importance and that, when things break, they need to be rapidly repaired in order to return them to service. A far different example of an organizational value is that equipment should never fail unexpectedly. In this case, the plant is reliability focused and are aware of problems as they develop as opposed to after they happen. These two examples are very different. However, in each case, the value described drives the collective decision making process for the organization and is very difficult to change.

What are your organization's values? Thinking about and identifying them may not be as easy as you think; the true values of a firm are not always written down. Instead, they reflect how the members of the firm collectively behave, how they conduct their business, and what they believe are the true measures of success.

For our discussion, organizational values can be defined as:

A company's basic, collectively understood, universally applied, and wholly accepted set of beliefs about how to behave within the context of the business. They also describe what achieving success feels like. These values are internalized by everyone in the company and, therefore, are the standard for accepted behavior.

When faced with a problem, those within the organization will invariably make a decision that reflects the organizational values of that business. These decisions are often not made consciously because organizational values are internalized and taken for granted. When you make a decision supported by the values, you feel comfortable. When you don't, you sense that something is just not right with your world.

Role Models

Role models are people within the company who perform in a fashion that the organization can and wants to emulate. They are successful individuals who stand out in the organization by performing in line with the corporate value system. Role models show people that, if you wish to be successful, you need to follow the values set up for the organization. These role models are then copied by those who work within the business because they show how to perform within the culture. In addition, the role models are used as an example for newcomers to clearly show how to behave if you wish to succeed.

Let us discuss further the three key components of a role model.

Top of the organization Most people who are used as role models are at or near the top of the organization's hierarchy. These are the people we view as the most successful. They are the managers of our departments, the leaders, the ones who set the direction for the business. The key word here is success. Because those at the top are perceived as successful, we tend to use them as role models.

There is another reason why we often choose our leaders as role models. They set the expectations of what we are to accomplish at work. In most cases, these expectations are in line with their expectations for themselves. As a result, we emulate and assume their style because we are all working towards the same end. In addition, failure to achieve these expectations usually has severe negative outcomes. Therefore, modeling the manager to achieve the desired results makes sense.

Successful within the organization's culture The second component is not only that role models are successful, but also they are successful within the existing culture. This is very important. Because role models are those whom we emulate and because they have shown that they can be successful in the existing culture, the existing culture is continuously reinforced.

A style we can identify with and adopt Even though some people are successful within the culture, there still may be reasons why we would not choose them as role models. If we truly want to use people as role models, we not only need to view them as successful, but also need to feel comfortable adopting their style of management. Suppose you are the type of person who firmly believes that all people within the workforce have unique value and should be treated with dignity and respect. Further suppose that your manager (who is a successful part of the organization) has achieved this position by acting and behaving in exactly the opposite fashion. Could you accept this person as a role model? Your answer would probably be no. Although you want to behave in a manner that will provide you a successful career, the behavior of your manager could never fit your personal beliefs and manner of conducting business.

The role model that is in conflict with our personal value system is worth further discussion. This type of person is the most difficult to work with. This is the person whose beliefs and actions are so opposed to our own that it is virtually impossible to adopt his or her style of management or behavior without violating who we are. There are alternatives when you are confronted with this type of situation. You can leave the organization and seek work elsewhere. You can attempt to transfer to another department. Or you can try to stick it out and survive, hoping that the individual will leave before you do.

Not everyone is a positive role model. We are often presented with what I will refer to as "good bad examples." These are people who we can look at and say "here is someone who I do not wish to act like." If you examine why you feel this way and adopt behaviors that are opposite and more in line with how you feel you should behave, then they will have done you a great service. They will have shown you a model that you will choose to reject for a more positive (and opposite) behavior when you become a role model later in your career.

Rites and Rituals   Rites and rituals are the work processes that go on day-to-day within a company. They are so ingrained in how people conduct business that they are not actually visible to those within the company. Rituals are "how things are done around here." Rites are a higher level of rituals. They are the events that reinforce the behavior demonstrated in the rituals. For example a work planning and scheduling process is a ritual. It denotes tasks that a planner performs each day as they do their job. Rituals are the reinforcing events such as a "job well done" by the manager for a planner who has developed a sound work plan for a critical job.

A ritual is a rule or set of rules that guide our day-to-day work behavior. Rituals are taken for granted because they are an integral part of what our jobs are and how we do them. As they are repeated daily, rituals become an accepted part of how business is conducted. Over time, they become invisible to those who follow them. Yet they are extremely important, not only because they define what we do and how we do it, but also because they represent our culture and the value system in place in our plant.

Furthermore, rituals are taught to new employees so that they understand "how work gets done around here." Rituals guide how people communicate and interact. Because they are so ingrained in our work, an outsider might say they were blindly followed—even if they made little or no sense. In addition, they are often defended fiercely simply because "that is how things are done." This viewpoint explains to some degree why new programs or processes that conflict with the plant rituals encounter such strong resistance when they are implemented.

Every one of us has had this experience. The first thing we are given on our first day of work is training in how the work is conducted—the rituals of the job or department and, what is more important, the culture in which we now reside. As a new employee, this training is highly important because we are being told not only how to act but also what is needed to be successful.

Many years ago, I received my first supervisory opportunity as a foreman at my plant. Before the foreman who I was replacing left for another area, we spent an entire week together. I learned how work was assigned to the workforce, how to interact with production, how to order materials, and many other tasks.

At the time, our plant was totally reactive in the way we conducted maintenance. When things broke down, our most important task was to repair them as quickly as possible and return them to service. Still I was surprised when after lunch on Friday the entire crew was not assigned additional work but stayed at their staging area. I questioned the foreman I was replacing and was informed that they were waiting for things to break so that they could rapidly respond to the problem, make the needed repair, and avoid weekend overtime. Being naïve, I asked why they couldn't be assigned jobs that could easily be interrupted. That way, we could get some work accomplished while, at the same time, be available to respond to plant emergencies. I was told in no uncertain terms not to "rock the boat" because "this was how things are done around here." This was the ritual followed by each foreman. The culture was not about to let me change it!

In our context, therefore, a ritual is an invisible day-to-day work practice that is accepted as how work is performed within the organization's culture. The ritual provides everyone with a foundation for how work is handled. Processes outside of the accepted rituals are considered alien. The organization will feel extreme discomfort when new rituals outside of the accepted norm are introduced, even though it may not know exactly why. When I suggested an alternate solution to waiting for things to break, I was reprimanded, even though the outcome would have been the same—we could have still responded to productions emergency needs.

A rite is a company ceremony or event that reinforces our rituals. In a sense, they provide a stage for those involved to dramatize the culture and organizational values to those in attendance. Rituals and rites go hand in hand because without accepted rituals, rites do not exist.

Rites can cover a large spectrum of an organization's events. They include performance reviews, training, conferences, service awards, and departmental and group meetings, all the way down to a pat on the back for a job well done.

Let us look at a simple example. Consider the foreman who kept his crew in their staging area on Friday afternoon waiting to respond to the emergency needs of production. Several rites were associated with this single ritual. The first of the rites is the "pat on the back." When production called, maintenance was available to make the quick fix. If successful, the foreman would get a pat on the back for a job well done—a rite positively reinforcing a plant ritual. This sort of success would eventually translate into another rite—a positive performance review, better salary, and a chance for promotion.

Conversely, if the ritual was not followed, the associated rite would have a severe negative connotation. In this case, production would complain about the foreman's performance, resulting in other potential problems for the foreman who was out of compliance. My idea of having the crews work on interruptible jobs on Friday afternoon not only violated a maintenance ritual, but also seriously threatened an established set of rites for the foremen—the pat on the back and others of more significance.

Cultural Infrastructure

The cultural infrastructure is the fourth part of the organizational culture model. This is the informal set of processes that work behind the scenes to pass information, spread gossip, and influence behavior of those within the company.

The various components of an organization's structure can be represented as blocks on an organization chart. Although the blocks each represent a function within the company, they can't stand alone. They need the connecting lines that tie them together, providing a linkage for all of the individual parts. This linkage is the cultural infrastructure.

For our discussion, we will focus on people and communications as the key elements of the cultural infrastructure. These components are the glue that binds the organizational culture together and promotes sustainability of the firm. Thus, our definition of cultural infrastructure is as follows:

Cultural infrastructure is the hidden hierarchy of people and communication processes that binds the organization to the culture and sustains it over time.

The cultural infrastructure includes:
- Story Tellers: promoting the culture through war stories
- Keepers of the Faith: mentors and protectors of the culture

- Whisperers: passers of information behind the scenes
- Gossips: the hidden day-to-day communication system
- Spies: passers of sensitive information to those who may or may not need to know
- Symbols: mechanisms for conveying what and who is important
- Language: terminology that describes what is done and often how

Table 5-1
Components of the Cultural Infrastructure

The Cultural Infrastructure		
Components	**What to Do**	**How to Do It**
Storytellers	Engage them in the change process	Get the storytellers to tell stories of the new way vs. that of the old.
Keepers of the Faith	Engage them in the change process	Have them provide counseling and mentoring, but focused on the new process.
Whisperers	Engage them in the change process	Work with them so that they filter and pass information supporting the change.
Gossips	Negate the value of their communication	Provide up front and continuous communication to reduce the material to gossip about.
Spies	Negate the value of their communication	Same as the gossips but it will undermine the value of the information they have to pass.
Symbols	Change to reflect the new reliability focus	Alter the symbols to reflect the new focus. Reward those who contribute to the new process.
Language	Change to reflect the new reliability focus	Provide education and involvement so that everyone speaks the same language.

Each of the cultural infrastructure components that we have listed above can be used to promote cultural change or, conversely, to disrupt it. Table 5-1 identifies each component, providing a brief indication of what and how you need to use them to successfully support your change initiatives.

Changing the cultural infrastructure is not an easy task. Great care and patience must be taken if you are going to make the attempt. However, you must understand that the cultural infrastructure is a hidden force that, if not dealt with, will most assuredly work to undermine whatever changes you are attempting to implement.

Cultural Change and Reliability

Figure 5-4 describes a reactive repair-based work scenario. First, something breaks down (Block 1). Then the problem is iden-

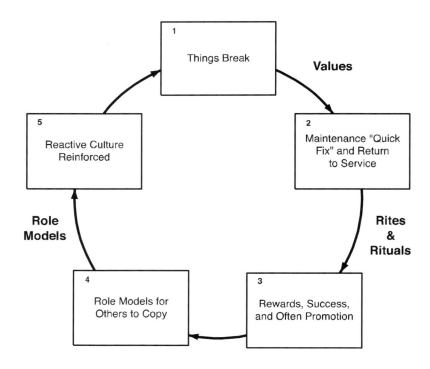

Figure 5-4   The Reactive Culture's Maintenance Model

tified; a maintenance crew dashes in, makes the save, and order is restored (Block 2). As a result, the crew receives praise from production for a job well done (Block 3). The praise is most often immediate and the reactive behavior is, therefore, immediately reinforced. Maintenance is now ready for the next emergency (Block 4).

As you can see, the organizational values dictate the response from the maintenance organization. In this case it is "drop what you are doing and fix the equipment that has broken down." In many cases, this type of response is required. However, all too often the quick fix or "emergency job" is not really an emergency at all—just someone's desire to get their job worked ahead of others. Nevertheless, when the call comes in, maintenance responds and is summarily rewarded. This response and the related reward are the rituals and their reinforcing rites at work. Over time, the rapid responders are rewarded for their efforts and become the role models for the organization.

What you see at work here is the perpetuation of the reactive maintenance model. Because we wish to change the culture to one focused on reliability, we need to alter the culture. To accomplish this we need to change the four elements of culture.

1. The organizational values must be altered. If the values support a reactive behavior, it is impossible to change the culture. This is the role of the leadership team.

2. Once the values have been altered, then work processes, structure, communication, and other basic operational processes must be changed. The rapid response can no longer be tolerated unless it truly is necessary. Additionally, a new reward structure must be put in place to provide those you wish to change with reinforcement for the new set of behaviors you wish them to learn. This is the way you can modify your rituals and their supporting rites.

3. The role models of the reactive process need to adopt the new reliability process or they need to be removed from role model positions. It is critical that you have the people in place to model the new behavior as it is being implemented. This will work on two levels. First the organization will see that, as work is executed, there are people in place showing them a new way that "things are going to be done around

here." Second, it proves to the organization that you are serious about the change.

4. Last you need to pay careful attention to the key members of the cultural infrastructure. Remember that these individuals are the behind-the-scenes communication network. You need their support, which often be gained by including them in the effort.

Just as with the eight elements of change, the four elements of culture are very important as independent elements. However, they are far more important when taken together as a set of dependent elements that must be addressed for successful change to take place. Figure 5-5 show what happens if any one of the four elements of culture is omitted from the change equation.

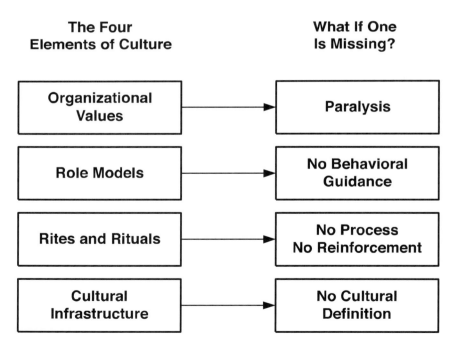

Figure 5-5
What if One of the Elements of Culture is Missing?

5.4 The Importance of Soft Skills and Culture to the Internal Consultant

Knowledge of the soft skills and the elements of organizational culture are critical success factors for the internal consultant. From the example in this chapter and, more important, from personal examples of your own, think about change initiatives that have failed. Then dissect the failure referencing the eight elements of change and the four elements of culture. If you conduct this exercise, you will quickly see where the change initiative broke down, setting the organization up for either short-term or long-term failure. Conversely, if you as the internal consultant can introduce and make certain that the soft skill and organizational culture elements are introduced at the inception of the initiative and then made an integral part of the process, there is a greater likelihood of success for the initiative and for you.

The introduction of these elements into the process is not going to be an easy task. Those involved and responsible for the change initiative are going to want to get the change started; they may have little interest in addressing the soft skills or the culture. In these instances, internal consultants have a critical role. They must convince the decision makers of the need because if these elements are not included, they certainly will cause the initiative's demise due to their absence.

Five Things to Think About or Do

1. Select an initiative that has failed in your plant. Write down the reasons why you believed it failed. Think about these reasons in reference to the eight elements of change. Can you see how missing elements contributed to the initiative's failure?
2. List the four elements of culture. Using the same initiative from question #1, identify the elements of the organization's culture that may have contributed to the failure. Can you see how the failure to address the cultural elements contributed to the failure?

3. Select a change initiative that was successful over the long term. Can you see where addressing the eight elements of change (consciously or sub-consciously) helped this initiative be successful?
4. Using the same initiative in question #4, can you identify how the four elements of culture contributed to the initiative's success?
5. Make a list of things that you can do as an internal consultant to make certain that change initiatives take into consideration both the soft skills (the eight elements of change) and the four elements of culture.

Learning Organizations

If you do the same old thing the same old way
You will always get the same old result
Group learning can correct this problem

6.1 Introduction to Learning

This chapter is about group learning specifically related to organizational change and even more specifically related to plant reliability. Every organization and those within the organization learn continuously. Although you and your organization may not learn the correct things nor apply what you have learned, nevertheless, learning goes on all of the time.

The questions we will address in this chapter are
1. What is group learning?
2. What constitutes a learning organization?
3. How do we acquire knowledge?
4. How does this apply to reliability?
5. How is learning supported or impaired by the organization's culture?
6. As internal consultants, what we can we do about it?

In our daily lives we learn something new everyday. It is important that we consider each of these new things that we learn and see how they are applied to the world in which we work. If we don't, there is a significant risk that the world will change and we will be left behind. Just think for a moment of the things you have learned over your career and how they have been applied towards making work easier and making you more productive. A simple example that has affected everyone is the computer. I can still remember doing things long hand and the large amount of time and effort it took. I can also remember the difficult time and extended effort required to justify one PC in our entire organization. Times change and we need to learn and apply what we have learned.

Although it is important that we learn and apply this learning, it is even more important for the success of our companies that we learn as a group—hence, group learning. The reasoning for this is the simple truth that no one individual can easily make a change for the better. You may have learned something that can vastly improve the reliability of your equipment. However, if that learning is not transferred to the entire group, it will never be implemented. The reasoning behind this is the fact that groups are what make up the organization and, ultimately, the company. If the group learns something new that will support their effectiveness and efficiency, and they are aligned in their approach, the result will be a powerful and successful change.

Group learning is an important organizational trait that has the potential to make the work of internal consultants far easier than if learning was not a key component of the organization's make-up. After all, internal consultants are responsible to work with the organization and develop new initiatives that will improve efficiency and effectiveness. If the organization is closed to learning and applying the actions required, that task will become difficult to impossible.

Years ago I came to understand that reactive maintenance was not nearly as effective as work that was well planned and executed. In one consulting effort in which I was engaged, I suggested that the organization move away from their reactive work processes towards those that had a much higher degree of proactivity. However, there were many in the organization who still embraced the old reactive way of working. Even with significant amounts of evidence they failed to embrace the new concepts. They were unable to learn that, even though they had been reactive for years, it was not an effective way of conducting the work. As a result, the change initiative failed. The ability of an organization to be able to learn and apply the learning to new and better ways of performance is not just critical to the success of change initiatives; it is also critical to organizational survival.

6.2 Group Learning Defined

An organization that will make a successful change from their current state to one focused on reliability and proactive maintenance needs to have group learning as one of its key elements of change.

To get a better understanding of the concept of group leaning, we need to define this term. In addition, we need to break down the definition into its component parts and develop an equally deep understanding of each part.

> Definition: Group learning is the ability of individuals to acquire new knowledge (about new or existing things) and employ this knowledge in a group setting, leading to aligned alternative courses of action or reinforcement that the current course of action is correct..

Let us take a moment and break this down so that we can discuss the significant parts and gain a better understanding of what group learning for an organization is all about.

6.3 Levels of Learning and Alignment

Learning in our organizations take place at three levels: the organizational level, the work group level, and the individual level. Each of these is a subset of the prior group. The work group is a subset of the organization and the individual is a subset of the work group. These relationships are shown graphically in Figure 6-1.

As you can see in the figure, individuals are part of work groups. Although we learn things at an individual level, the learning never really takes root unless the idea is embraced by the team of which we are a part. Later we shall learn how the culture of the organization and the working groups (sub-cultures) influence learning.

The work group level is where learning can gain a foothold. However, this is not as simple as we often wish it would be. Work

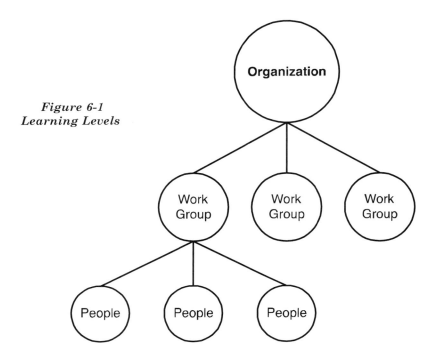

Figure 6-1
Learning Levels

groups are dynamic. People learn things in different ways and draw conclusions that may be different from other members, even though they are presented with the same basic information.

Suppose you attended a seminar where you learned about a new technique for aligning rotating equipment. From this experience, you recognized that this new procedure could significantly improve the effectiveness of the mechanics in your plant. However, when you introduced the concept to your work group, the idea was met by resistance by many of the members. Some felt that it would diminish their skills. Others felt that the new process could be handled by just a few experts and would ultimately cost them their jobs. Only a few liked the idea and could see the benefits.

As you can see, a simple new idea that would improve how work gets done met with various reactions from the group—many against the idea. This misalignment of what people understand about the information that you provide can seriously derail even the best of ideas. It can be overcome by slowly introducing new ideas and handling them on a pilot basis. This gives everyone the ability to see the benefits and understand the impacts. Most often when this approach is taken, successful implementation of a new idea is made easier because people have time to accept it.

If it is hard to achieve alignment in a group, just imagine how hard it will be to take learning from one group and apply it across an organization.

6.4 How We Acquire Knowledge

We acquire new knowledge from both external and internal sources. Externally we encounter new information all of the time. We learn from attending conferences and seminars, meeting with other companies and their representatives on industry committees, meeting and working with external consultants, attending training classes, and many others. Internally we learn from our constant interaction with other plant sites, other departments within our own site, and new hires who have a much more objective view of what we do than we do—at least for the first six months of their new careers.

The problem we face as we learn new things in each of the above cases is that we continually filter information. Learning new things is one step towards change. However, the harder step is being open enough to allow these new ideas past our filters so that we can analyze the information and determine if it is applicable. We need to be open to the fact that someone else may have invented a better mouse trap. If we fail to turn off the filters and learn, we are doing a disservice to our company by potentially missing out on new ways to improve the work. Filters of this type look and sound like:

- That is not how we have done this work before.

- We have been successful with the old method; why change?

- We tried that before and it was a waste of time.

- Our way of doing _____ is better.

- That is a different industry—it won't work in ours.

- That can't be right our information says something different.

There are many others that could be added to the list. Each in its own right is a destructive force that can render the best new idea useless. To be successful learners, individuals and groups need to turn off the filters and be open to new ideas. Rather than dismissing them, analyze their value. If there is value, use the idea to improve the business. If there is no value, then dismiss the idea, but at least you will know that its potential was evaluated first. The learning organization has the ability to recognize the gaps within their own sphere of knowledge, then to create or acquire new knowledge to fill these gaps,

6.5 How We Employ This Knowledge

Figure 6-2 shows a single loop where we set goals, act on the goals, and compare the results of our actions. Where there is a gap between our goal and the experienced outcome from the work, we identify it and feed it back into the work system. This causes us to readjust our work activities so that our next outcome more closely matches the desired goal.

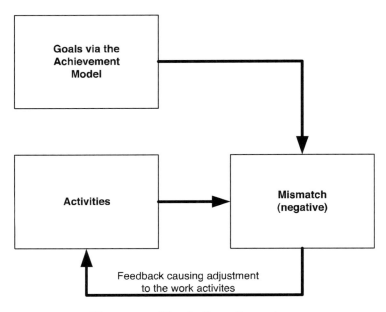

Figure 6-2 Single Loop Learning

Eventually, through continual application of the learning loop, we will have closed the gap and reached the goal. At this point (or even before it), we may want to readjust the goal to something different. This new goal, which is most often more challenging than the one prior to it, is often referred to as "raising the bar."

Suppose we want 100% compliance with the preventive maintenance program (PM) so that PM tasks are completed as scheduled. Our goal, therefore, is 100%. We then conduct our program and measure the results. Assume that the first time we do this our compliance is at 80%. This result is then fed back into the work process; changes are made that will affect the rate of compliance. Assume that the outcome of the second iteration is now 90%. Once again, we review the work outcome and adjust the process until at the fourth iteration we reach the goal. Figure 6-3 demonstrates this process graphically.

This single-loop learning process is relatively simple to apply to single activities such as a PM program. But think how much more difficult it becomes when applied to the many and inter-related goals of the business.

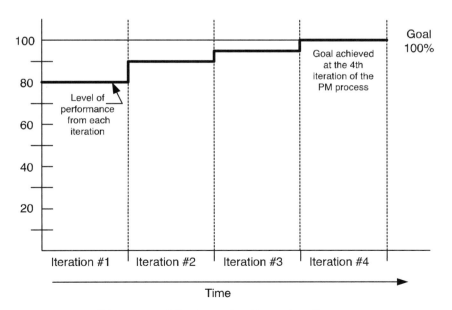

Figure 6-3 The Iterative Learning Process

There is also a flaw in the single-loop learning process. What if the goal against which you are comparing your activity outcomes is wrong? Not possible, you say! However, think for a moment about organizations that have rapid repair or "break it–fix it' as their mode of operation. Their goals are then centered on improvement to the repair process. These organizations have feedback loops telling them to strive to fix things faster. We already know that in most instances this goal is expensive and inefficient. What can be done to break this flawed feedback process? The answer is the employment of double loop learning, as shown in Figure 6-4.

In this model, there is an additional process. As with single loop learning, we examine the outcome of our activities vs. our goal and make needed adjustments to our activities. But in double loop learning, we also re-examine the goals themselves and make appropriate adjustments.

Let us go back to our PM example. Suppose that when we reached the 90% compliance level, we noticed the benefits from

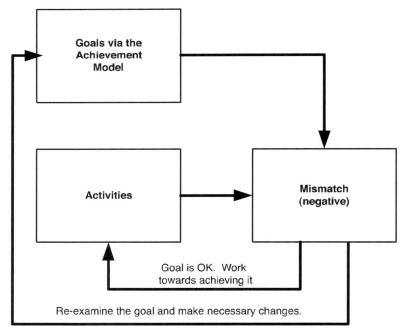

Figure 6-4 The Double-Loop Learning Process

attempting to reach 100% were minimal as opposed to the amount of effort, time, and money that would be required to achieve 100% compliance. Do we want to continue chasing this goal? The answer may be no. Instead we need to re-examine the goal and make adjustments at a higher level. Maybe our goal needs to be changed to 90% PM compliance and 10% predictive maintenance. (For example, we could decide not to worry about PM on equipment that doesn't need it.). This shift brings into play a different feedback loop—not working to alter our activities but to alter our goal.

Understanding and employing this process is often new, different, and difficult for an organization. However, as you learn you evolve; the things that you set as goals at the beginning need to be altered based on your learning. We will talk more about how this works when I address learning spirals. Remember: "if you do the same things always in the same way, you can never expect to get anything other than the same old results."

The next question that needs to be asked is when do you re-examine the goals? The rule of thumb is that goals should be re-examined when the current level of activity is no longer generating incremental benefit. This is often different for different organizations and is not usually tied to a calendar, even though companies seem to feel that goal setting is an annual event.

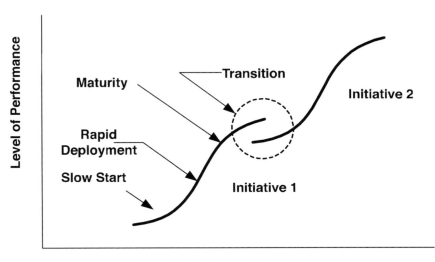

Figure 6-5 Initiative Transition

Goals, their review, and often readjustment fit what is referred to as the S-curve model. Figure 6-5 shows how this works. In this model, the x-axis represents time and the y-axis is the level of performance from low to high. The activity which we are tracking against our goal starts off slowly. As time progresses, it is accepted; as performance feedback is fed back into the process, performance begins to rapidly improve.

Then the initiative reaches maturity and progress slows down, often reaching a plateau where addition improvement is non-existent. It is at this point that the second loop of the double-loop learning model needs to be employed. As the existing goal is reviewed, we recognize that it is time to set a new goal that once again will add significant value to the company.

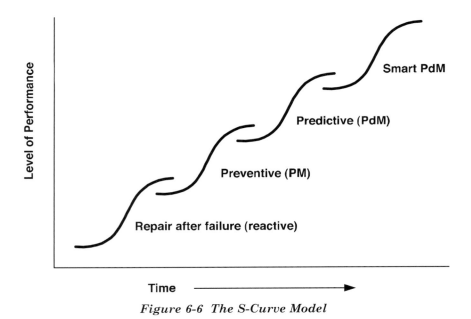

Figure 6-6 The S-Curve Model

Figure 6-6 shows the S-curve model for maintenance in some companies. In this model, the company started off in a reactive mode of operation. Over time they learned that this way of working is both inefficient and ineffective. They then evolved to a preventive maintenance (PM) focus similar to the example presented earlier in this chapter. After a while, this too failed to deliver the value that they sought and they once again evolved to predictive maintenance and so on through continued feedback loops that alter their goals.

6.6 Spiral Learning

The single and double loop learning models provide us with a valuable insight into the process of learning. Namely the process of learning and employing new things is not linear. Learning does not proceed as would a work project in which we know what the outcome is going to be before we start the work. Learning is far different. Learning takes place in spirals—in a manner similar to the single and double loop models. as shown in Figure 6-7.

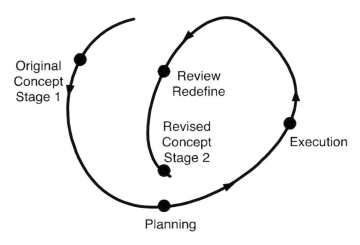

Figure 6-7 Spiral Learning

In spiral learning we identify a new concept or idea, plan our actions, execute according to our plan, and then review the outcome (using the feedback loop models). This review will then direct us to alter not only our activities, but even our goals. Realize that when we start any loop we often do not know what the second, third, and subsequent loops will look like until we get closer and have learned from our actions. This point is especially important for engineers and their managers, many of whom have been trained to think linearly. Linear thinking does not work well in the world of change where activities, results, and future activities are non-linear. For a more detailed review of this subject, see *Successfully Managing Change in Organizations: A Users Guide—* Chapter 3.

6.7 Learning and Blame

As we perform various activities in an attempt to achieve our goals, things often go wrong. The wrong procedure is applied to the work, the wrong part is installed, the equipment is started up incorrectly, and a whole host of other things that affect production as well as the performance of the workforce. Someone once said that if things can go wrong, they will; no matter how hard we try to prevent this from happening, it still does.

Learning organizations will approach these failures far differently than organizations which I will call blaming organizations. In learning organizations, the leadership will say, "What can we learn about this problem? We want to put processes or procedures into place so that we can avoid it ever happening again." On the other hand, blaming organizations will search for the guilty, then assign blame and punishment. In the former approach, improvement is a definite possibility. In the latter, a repeat of the problem is a certainty.

Learning organizations will work very hard to search out the root cause of the problem. By correcting the root cause, it will eliminate the potential for reoccurrence. This can never be accomplished if blame and punishment are part of the equation. The reason for this is that people will not tell the truth about the incident for fear of being blamed and punished. Why would a mechanic or production operator in a blaming organization admit to making a mistake if the consequences could be avoided by playing dumb? So they say nothing. Meanwhile, the real problem—that the wrong part was in the warehouse or that the procedure for start up was wrong—goes unnoticed and we don't learn from our failures.

Learning organizations address each of these problems differently. They provide amnesty for these individuals; they guarantee that there will be no punishment if during the investigation they reveal what happened. In this way, learning organizations can get at the root cause of the problem and prevent reoccurrence. However, an amnesty policy sets up a conflict. How do we provide amnesty and still hold people accountable for their actions?

There are two types of actions that need to be considered if we are to answer this question. The first type includes inadvertent actions—things that we did because we believed that they were correct. The second includes willful actions—things that we did even though we knew at the time that what we were doing was wrong.

Inadvertent Actions

When inadvertent action causes failure, blaming the individual that caused the actual failure is useless. They are often the last one in a long string of failures that led up to the event. Take, for example, the wrong part in the warehouse. Is the person who installed it after taking the number from the bill of material to be blamed? No. The cause of the failure needs to be spread far across the organization. Part of the cause was the engineer who wrote down the wrong part, the purchasing agent who bought it, the vendor who should not have supplied it but did, the rotating equipment engineer who told the mechanic to install it, and finally the mechanic. Learning and ultimate corrective action is far better served by making certain that everyone understands that there is no blame and punishment for this type of failure. In this way, through amnesty, we will learn the root cause of the problem. The goal is that the wrong part will never be installed again.

Willful Actions

The second type of action—willful action—is far different. In these cases, people do the wrong thing even though they know it is wrong. They are breaking what are referred to as the laws of the land. These actions do bring blame and punishment. For example, mechanics who bypass a plant safety rule, exposing themselves and their co-workers to serious injury, have broken a law of the land; they deserve the blame and punishment that accompanies it.

The important thing for a learning organization to provide is a clear understanding of what these laws are at the work site. People then will be able to differentiate easily between inadvertent actions and willful actions. The amnesty rules can be applied to improve our learning from our failures.

6.8 Learning and the Soft Skills: The Eight Elements of Change

It is easy to see how learning can be applied to the other seven elements of change. Every one of these and the actions we take that relate to them are tied to both single loop and, quite often, double loop learning as we develop and improve upon our goals. Without this element of change being active, we are doomed to repeat our mistakes, yet never really understand why.

There is a quotation that immediately comes to mind when thinking about this issue. "If you do the same old things in the same old way, you will always get the same old results." Isn't this exactly what happens when we fail to learn and apply what we have learned? Everyone needs to keep their eyes and their mind open to new ideas and how these new ideas can be applied. The eight elements of change provide a framework in which to conduct this learning examination. For example,

- **Work Process.** Recognition that a reactive maintenance work process is both inefficient and ineffective, then developing a proactive process to improve equipment reliability.
- **Structure.** Identifying the fact that the current work structure may not support your longer-term reliability goals, then changing it.
- **Technology.** Determining that the systems being used in support of the business may be antiquated, then finding and implementing new ones.
- **Communication.** Understanding why communications within your plant never seems to get the message across to the workforce, then changing the process.
- **Interrelationships.** Learning how the various interrelationships within your business either support or fail to

support the new change initiatives, then doing something about the problems.

- **Rewards.** Recognition that the old reward structure may not serve to promote change. then altering it to one that does.

6. 9 Learning and the Organizational Culture

As we are learning and making changes to improve, there is also an aspect of learning going on within the culture of the organization. All throughout our discussion, we have talked about changing the culture from one focused on reactive equipment repair to a new culture where equipment failure is not accepted as the norm. As we bring in new managers, change the work processes and structure, implement new computer systems, and work through other initiatives designed to accomplish this cultural shift, the organization is watching and measuring the effort.

The existing culture—whether good or bad for the business—is one that has been around for a long time. Many of the managers and others who have had successful careers have achieved this by working within the current work culture. As a result, the culture and those within it are going to be very skeptical of a new way of doing things. However, new ways can be learned and new processes adapted if the culture can be convinced it is within their best interest.

Every one of us has been exposed to change programs throughout our career. Most of these have fit the model of "program of the day." Over time or because of managerial changes, these programs have come and gone. This model is what people expect for every program of this sort. Yet if we are to be successful, we must change this belief as a first step in the process. This is accomplished by paying careful attention to what we are doing and how it affects the four elements of culture. Figure 6-8 depicts the relationship of the four elements and shows how they actually work as part of a system. Failure to address these elements and their interrelationship can spell disaster for change initiatives. Yet if we address them properly, the organization will learn that we are serious and we can shift the culture.

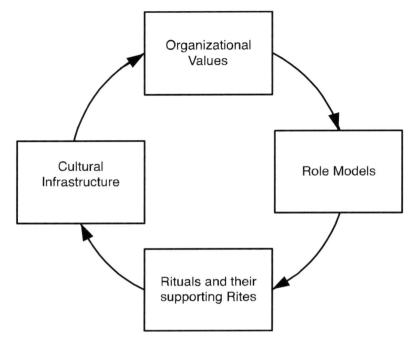

Figure 6-8 The Four Elements of Culture System

Organizational Values

The values of the organization are paramount. These values are often expressed as a mission statement, written down and widely distributed to the organization. If you do only this, you fall far short of the expectations of the culture and what they need if they are to embrace the new way of working. They want to see management "walk the talk." They want to see action in support of the values written down. They want to see it all of the time and they want to see it consistently applied. It takes a great deal of effort to establish this level of cultural credibility, but this is how the organization learns that you are serious. Examples include:

- Proving that preventive maintenance is really important by not reassigning the crew—ever!
- Learning from failure by trying to understand the reasons and how to prevent the failure in the future; it not through punishment of those who seemingly failed.
- Telling people training is important and providing it.

- Praising the proactive work efforts far in excess of those reactive work efforts.
- Hiring people with skills that support the new process.
- Investing in better reliability. Cutting costs while looking good in the short term spells long-term problems while undermining the organization's confidence.

Role Models

Once the value system is in place, the next step in the learning process is to provide the role models that exhibit the way you wish to work. This is at times painful while those who are the role models for the old way of working get moved aside. However, if this is not done, the organization will quickly learn that although the new values have been established, there is no teeth in the message. The evidence for the organization is that the old role models are still running the show in the old way.

Rite and Rituals

With the values in place and the role models established to support them, the next thing that must be altered are the rites and their supporting rituals. This is the third step in the learning process. This step must be completed or there will be major disconnects between expected performance and what is actually taking place. This is the third step in the organizational learning process.

The Cultural Infrastructure

Last but not least is the cultural infrastructure. The infrastructure is anchored in the old way of working. However, if the first three steps are handled properly, those with primary roles within the cultural infrastructure will learn new things and change over time.

6.10 What Does This Mean for the Internal Consultant?

Organizational learning skills and the recognition by internal consultants of the level of organizational maturity to learn are both critical success factors for any change initiative. They are critical for both the organization as well as for you on a personal level. Imagine trying to work with an organization to implement a new process if the organization is not open to learning and applying new ideas. How successful do you think you would be in this scenario? How successful do you think you would be if the organization essentially refused to adopt new ideas? If it blamed individuals for failure? If it didn't recognize the negative effect that not learning new ideas had on the soft skills and the organization's culture? The answer is obvious; the initiative in which you were involved would most likely fail. In fact, you probably would be blamed for the failure.

As an internal consultant, what can you do to correct this problem? You can try to educate the organization to the value of learning and applying new skills. Examples of group learning and its application to the process of change may help the leadership see the light. However, the things that make an organization one that refuses or resists learning are often rooted in how they apply the soft skills and also in the way that the organizational culture has been established and perpetuated. These are things that, as an internal consultant, you will find very difficult to change without site support. You can not change them on your own!

Your best bet is to make certain that you have clients that will support the learning process and help the organization make the transition. Otherwise, you should not get involved in initiatives with groups that refuse to learn. However, as an internal consultant, you often do not have this luxury. When you find yourself in this position, you need to get support to help you create a learning organization where it may not have existed before. You also need support to help prepare the site for a great deal of frustration and a high likelihood of failure of the initiative if they refuse to embrace group learning as a part of their change process.

Five Things to Think About or Do

1. Make a checklist for yourself. List the traits you would look for in an organization to determine if it is a learning organization.
2. Review you list of traits against the way that your organization functions. Is your organization a learning organization or not? What traits are missing? If it is a learning organization, write out several specific examples that make your case. If not, what traits are missing and how can you help your organization acquire them?
3. Does your organization fit the blaming model when things happen that negatively impact the business? If this is the case, write out several examples. What would need to be changed to create a true learning organization to replace the current model?
4. Think of an initiative that has been successful. How did the element of group learning as it relates to the eight elements of change and the four elements of culture support the success?
5. Think of an initiative that was not successful. How did the lack of group learning related to the eight elements of change and the four elements of culture help the initiative to fail?

CLIENTS

Clients are the internal consultant's nourishment
You need them to survive and even more
to be able to grow

7.1 The Need for a Client

Imagine that, as the site's internal consultant, you recognize that your organization needs an improved process to plan and execute outages of the major production lines within your plant. You have come to this conclusion by observing first hand the poor development of the work plan and execution of the work during the last outage. You recognize the severe problems that these deficiencies have caused such as failure to meet the scheduled completion date, overrunning the cost estimate, and completing the job without the expected level of quality. In response, you spend a great deal of time researching how others plan and execute the work, attending conferences, and talking with industry experts. You then develop a very comprehensive plan for conducting the work in a more effective and efficient manner. This process has proven to be a successful technique throughout industry; you are firmly convinced that it will do the same for your company.

At the presentation of your recommendations, you are surprised. Although the managers appear to be listening, they don't seem to have any real interest in implementing what you have developed. You leave the meeting discouraged; your presentation along with the improved work process ends up in your file. The next outage is as chaotic as those preceding it. Without a clear pre-planned work process, the job is not completed on time and is over budget. Your plan would have altered this outcome. Something has happened that is detrimental to both you and the business.

Now let's consider the same scenario, but with a twist. In this case, when you recognized the need to develop a detailed work process for the major equipment outage, you discussed this with the maintenance manager, who assigned you the task to develop the process. An alternative to this could be that your managers recognized the need on their own and assigned you the task to develop the process. As a result, when you developed your recommendation and made your presentation to the manager's staff, the process was accepted and implemented.

There is a key difference between the scenarios. In the first, you did not have a client within the business unit who supported the work that was done. As a result, when it was presented, there was no buy-in by the organization and, hence, it was rejected. We have learned in earlier chapters that the internal consultant's role is to support the work of those who are responsible and accountable for its execution. Although people in the internal consultant's role may clearly see the business need for specific change initiatives, those initiatives will not be adopted unless the work is conducted at the request of a business client.

7.2 Who Are Your Clients?

As you can surmise from the example and probably from your own personal experiences, you need a client in order for internal consultant-supported efforts to even be considered for implementation. Clients can vary from senior managers all the way down through the organization's hierarchy. However, they have one key thing in common—the work that you do for them must have the ability to be implemented.

Suppose a senior manager asks you to support a work redesign effort. It is highly likely with a client at this level that the work, when completed, will be implemented. This could also apply to a departmental manager or even someone at a lower level in the organization. I define a client as:

Any person or group within the organization who has the ability to contract with the internal consultant for work which, at its conclusion, has associated with it the likelihood of implementation.

Let's dissect this definition into its component parts in order that we gain a better understanding of the concept of the internal consultant client.

- Any person or group who has the ability to contract
  It is true that any person or group can contact the internal consultants requesting that they take on an internal consulting contract. But not everyone has the ability to finalize the contract. This is a good thing for both the internal consultants and the management team.

- Suppose that the internal consultants were asked to support an initiative that was conceived at the foreman or worker level. Although they could take on this work, there may be little support for the outcome. As a result, there could be hard feelings among those who asked that the work be conducted. In addition, the credibility of the internal consultants could be undermined since they took on an assignment that was not supported by the management team.

- On the other hand, suppose that the work was requested by the senior management team. In this case, the internal consultants would have support for the time and effort that they and others would have to put into the initiative. The important point here is that, while anyone can contact the internal consultants for work, only certain people within the organization have the authority to do so.

- At its conclusion has the likelihood to be implemented
  The second part of the definition is equally important. A client is not just someone who has the authority to request the services of the internal consultants. A true client is in an environment where there is a high likelihood that the work will be implemented. This does not imply that every initiative, once it is developed and reaches the "go – no go" stage. will be implemented. But someone who has the potential to be a client should at least be able to state that there is likelihood of implementation once finished. If this were not the case, why would senior management, the group who created the position of internal consultant, be willing to keep funding the efforts of the incumbent?

Clients, therefore, require two very important traits. First, they have to have the authority to assign the internal consultants to a specific initiative. Second, they have to have the ability to make certain that the work, once completed, is highly likely to be implemented.

7.3 Client Expectations from an Internal Consultant

When a client contracts with internal consultants, there are certain things that they have a right to expect as a product of the contract. Obviously the first and foremost is a successful work initiative supported by the internal consultants and owned by those on the site team. However, there is more that a client expects from the internal consultants on a more personal and professional level. These include the following.

Skilled and Timely Workmanship Throughout the Work Effort.
The internal consultants are expected to have the skills needed for consultant-type work as well as being able to apply these skills to a timely delivery of the work product. After all, being able to conduct a skilled effort is worthless if it is delivered late.

Understanding of the Scope.
At the outset of the work effort, there is agreement on the scope between the client and the internal consultants. It is the internal consultants' responsibility to make certain that the scope is fully understood so that what has been requested will be delivered. This may take some questioning; often the full scope is not always clear in the client's mind. Nevertheless, the internal consultants must make certain that the scope is understood before the work begins. One good approach to this is to write out and then state back to the client what they have requested in the scope. The writing part forces the internal consultants to think about the work content. The restating of the scope back to the client gives them a chance to feel secure in the fact that the internal consultants understood the requirements.

Communication and Validation to Keep the Effort on Track.

At the outset, there is a scope of work developed which will be discussed in detail in Chapter 8. However, as the initiative evolves, there is a danger that it will get off track or that additional scope will be added. In either of these cases, the end result—even though it may have value—may not be what the client actually wanted. Therefore, it is incumbent upon the internal consultant to continually stay in communication with the client and periodically validate the direction.

Authentic and Honest Responses and Feedback.

One of the main reasons that the internal consultants were contracted was that the client has trust in their ability to deliver the work product. Not only that, but the client needs to believe that the internal consultants will be entirely honest in their communication and feedback. There is nothing worse or more damaging to the credibility (and longevity) of internal consultants than to provide feedback that is not authentic, honest, and complete. This means not only telling the client the good things, but also telling them the bad. Suppose an initiative which looked great on paper emerged as something that, if deployed, would cause more problems than benefits. Although providing advice that would terminate the effort may leave the internal consultants without an assignment, it is absolutely essential that they make this communication to the client.

Confidentiality.

Many of the initiatives on which internal consultants will work are confidential in nature. There are many reasons why the client may not to want the information leaked to those in the plant or even possibly those in the general public. The internal consultants must at all time maintain this level of client confidentiality if they wish to maintain the professional standing in the plant, their credibility with the current and future clients, and often their very jobs.

Balance.

The internal consultant works between two extremes. On the one hand, there are initiatives where they are expected to do all of the work. This is a poor expectation on the part of the client; by doing this, there is not ownership of the effort as it is developed or, more important, after it ends. On the other hand are the internal consultants who feel that they should not do any of the work. They believe that their role is strictly advisory in nature and that all of the work should be done by the team. This view is also poor because the consultants are brought into the effort to provide their skill and guidance—traits which those on the project team may not have. Therefore, the internal consultants need to work in a balanced manner—not doing too much of the work and not falling into the role of advisor only. The client expects the internal consultants to understand this and work in balanced manner.

Exit Ability with Ownership in the Customer.
   The client expects that every effort has some point at which the internal consultants need to exit and leave the day-to-day work to the organization. This is an expectation which the internal consultants need to appreciate from the beginning of the job. The simple truth is that as the job starts, the internal consultants need to be preparing for the site personnel to take over and for them to exit.
   If this does not occur, then one of two things happens. First the internal consultants no longer are acting in an internal consultant role. Instead, by failing to exit, they have become part of the organization conducting the work. Second is the problem where the site personnel are so dependent on the internal consultants that when they finally do leave the work product completely fails because the internal consultants were the ones actually making it function.

7.4 Internal Consultant Expectations from a Client

   Just as a client has certain expectations of internal consultants, the internal consultants also should have certain expectation of the client for whom they are working. These are as follows:

A Clear Scope of Work.
   It is difficult for internal consultants to deliver value if the scope of the work is not clearly defined. In these cases, the interpretation of the requirements of the effort are left to the internal consultants and the work team. Unfortunately, none of the members of

the team know how to read minds. As a result, it is highly unlikely that they will deliver exactly what the client wishes but has not clearly articulated.

This can be a serious problem. It is up to the client, who has a general sense of what they want, to be able to work with the internal consultants to develop a scope of work which can be utilized to drive the initiative. This may take time and several meetings, but a clear scope is an essential part of any successful initiative.

Leadership.

Once the scope of work has been finalized and the work has started, the internal consultants should expect that the client provides a certain amount of leadership to the effort. This does not mean that they become involved in the actual effort; although there are times when this is a requirement. However, it does mean that as the initiative evolves, the client participates in a way that the organization clearly sees that this work is important to the client (usually a site manager) and the company.

Sponsorship.

The internal consultants also should expect that the client provide ongoing sponsorship for the initiative. What this means is that the client, who typically is a part of the senior management team, visibly supports and acts as an advocate for the effort as the work is taking place. Without sponsorship, the collective focus of the organization will shift to alternative initiatives over the time it takes for the internal consultant to complete the work. As a result, when it is time to deploy what has been created, the focus of the organization will be elsewhere, making this task difficult if not impossible. It is the job of the sponsor to keep this from happening.

Motivation.

Longer change initiatives go through a life cycle. At first there is great enthusiasm. As the work progresses and people see how long the effort is going to take, there is a decrease in energy. This is especially true if the team members working on the effort have been assigned to work on it in addition to their regular jobs. It is the client's responsibility to help motivate the team. The internal consultants can accomplish some of this, but motivation from the client is far more significant and empowering than that provided from a third party. While this is an expectation of the client, the internal consultants often may need to ask for a motivational effort. The reason for this is the client is not closely connected with the work taking place and may not know when this is actually needed.

Commitment.
The internal consultants also should expect the client to be committed to the initiative over the long haul; the usual amount of time it takes for change efforts to actually deliver the value for which they are developed. All too often, management's focus is diverted to other issues; they cease to provide the commitment that is needed for a change effort to succeed. There are indicators that this is occurring and the internal consultants must be paying attention. When they see these signs, they need to act. This can be difficult because it falls to the internal consultants to remind management of the commitment that they made to the effort. Failure to sustain commitment is easily recognized by the team. The project will fall apart if the team believes that there is no commitment to its success from the client.

Funding and Resources.
Another expectation that the internal consultants should have is that the client will actively work to obtain needed resources in both funding and the more difficult resource to obtain: people. Internal consultants have no authority to provide either of these critical resources for the project. However, the client does have this authority and needs to exercise it in support of the project.

Recognition of the Time Required.
All too often the problem with a change initiative is that, once management sees the benefits that the effort can deliver, they want it implemented right away. This is not only a mistake. It is also an impossibility if the site management want the effort to be delivered correctly and have it generate long-term value. Internal consultants expect that the client will work diligently to convince the managers that, in the world of change initiatives, the proper time needs to be allowed to do the work correctly. In other words "slow is actually fast." Convincing others on the management team of this fact is critical for success of the initiative.

Ownership of the End Product.

One important task for internal consultants is the ability to disengage from the work after it has been completed. It is very important that this disengagement happens and that it is begun early in the actual work. Then, when the work is finished, it is owned by the client and their organization. The internal consultants have the responsibility to ensure that the client fulfills this responsibility so that the initiative's longevity is not contingent on the internal consultants remaining on the job. Otherwise, even if the effort is a success, it still has failed to achieve the true value built into the design.

7.5 Things Not to Be Expected

There is one thing that internal consultants should not expect—that their advice or recommendation will always be accepted and applied. All too often, internal consultants work diligently for a client and arrive at what they feel is a set of recommendations that clearly will provide value and address the organizational issues that they were hired to correct. Yet after the presentation, the management team rejects the proposal.

This is very frustrating for those of us in the internal consultant role. Often when this happens, we ask ourselves, "What have we done wrong?" Why, with all of our efforts and sound reasoning, did the proposal get rejected? Could we have made our points differently? Better? In a different format? To a more informed audience? ...or any of a thousand questions about our abilities. The truth is that the fault is not ours. Additionally it is not the client's fault either. There are many reasons why the proposal may not be accepted. Most of them lie either in the soft skills or in the organization's culture, neither of which the internal consultant has any control over.

At one point in my career, I led a project to install a computerized maintenance management system in all of our plants. This single system replaced computerized systems in each of the plants that were different from all of the others. This project lasted four years. At the end of this project, one of the suggestions that I made in my internal consultant role was that we set up a users group. This group would have representatives from each of the sites. Their role would be 1) to interact with each other, sharing the best practices for the use of the system, and 2) to interact with the software vendor, promoting enhancements that would improve the functionality of the application. The benefits of this user group were apparent to me and many others. However, when we made our presentation to management, the idea was dismissed and the user group never formed. At the time, I blamed myself for the failure to get the managers to accept what I truly believed was a very good idea.

The failure to get the internal consultant's advice taken is not the fault of the internal consultant. So let's take a step back in time and analyze what happened from the standpoint of the soft skills, the eight elements of change. (Note: This analysis focuses on the elements most relevant to this particular situation.)

- **Leadership.**   The senior manager believed that all you needed to do was deploy a system and the work process changes would follow. He was the one who tossed the project team out of his staff meeting for suggesting a work process redesign as a part of the software deployment. With this mind set, why would he or his staff consider a user group?
- **Work Process.**   The leadership team did not see the connection between work process and software. With this being the case, why would they want a group working to develop best work practices around the software application?
- **Group Learning.** The belief was that deployment of the software was the project. There was no expectation that there was anything to learn after the software was deployed. With this belief, the team felt there were more important issues to be addressed than to establish a user group.
- **Technology.** The view of this element was that once the software was deployed, there was no need to do anything further. They did not have any concept of improved functionality and software enhancements simply because these did not exist in their current systems.
- **Communication and Interrelationships.** A user group requires both of these elements, but the plants did not operate in this manner. The supporting evidence of this was

the fact that each site previously had its own maintenance computer system with little similarity between them.

The four elements of culture also had a hand in the failure of the management team to accept the idea of a systems user group.

- **Organizational Values.** The values of the sites were of independence, not cooperation to develop best practices and share information. With this mindset, the recommendation to form a user group was doomed from the outset.
- **Role Models.** The role model in this case was the senior manager. His prior behavior regarding work process change made it were clear to his subordinates what he thought of systems applications for anything other than support of the existing work process.

With all of these factors against the concept of a user group, it was no wonder the idea failed to get accepted, let alone implemented. This evidence also clearly points out that there was no inherent failure on my part. The cards were stacked against me before I started; I just didn't know it at the time.

The message here is simple. When a proposal is rejected, take a step back and ask yourself, "Did the team that I supported as an internal consultant do the best job we could have?" If the answer is yes, then review the rejection in terms of the eight elements of change and the four elements of culture. You may be surprised at what you learn about why your proposal failed.

One further thought for your consideration. Five years later with a different management team in place and a different set of values, I made the same proposal and it was accepted. All that had changed was the date on the documentation and, most importantly, the clients.

7.6 How to Get and Keep Clients

If you are in the role of an internal consultant at your site, you most likely did not get there by chance. Instead you probably delivered several special projects working outside of the day-to-day work process; the management team then recognized the value of having you in this position. Because the end point of every initiative is disengaging yourself from the work so that the true owners can take over, you need to continually keep a backlog of work so that you have things to do. The alternative is that you run out of work, no one requests you to work on new initiatives, and you wind up back in the day-to-day work process. This serves no value to you as the internal consultant or to the company.

The question is how do you get and keep clients? This is not as easy a task as that of the external consultant who has a far broader potential client base. The external consultant can get work from numerous companies across the entire country or maybe even around the world. On the other hand, you can only acquire additional work assignments from within your own company. Although the number of potential sites may be large, it is not as large as the opportunities available to the external counterpart.

There are three ways to obtain clients. First is to do really good work for your existing clients and have them tell others about the internal consultant skills you can bring to a change initiative. A second way is to keep very aware of initiatives being discussed and then proactively offering to get involved; a third is to recognize problems that are negatively impacting management and making suggestions to correct them.

In the first method, you have very little control of whether you get a lot of work or very little work—you are completely dependent that your name and what you can deliver are being spread by word-of-mouth by others. This isn't the best way to stay in the internal consultant business. It is far better to focus on the second and third approaches. In the second, you maintain an awareness of the initiatives being identified by others, then offer your expertise to help the organization turn these into successful change efforts. This is a sound approach because invariably initiatives identified in this manner already have sponsors. This makes your job easier; although sponsors have the ideas, they need you to carry them to fruition. The third approach is similar to the second, except in this case you are the one identifying the initiative. Unlike the second approach, there is no site sponsor, so you have to find one; otherwise, your good idea may very well go unrealized.

One more thing to think about. After you have a client, how do you keep them so that you get additional work? The obvious answer is that you continually deliver work of the highest quality, going the extra mile whenever it is required, and some times when it is not. Once you have established this high level of performance, an existing client will likely assign you additional work. The confidence level that high-quality work generates may also see the client discussing their ideas for the future with you, which also may help you get additional work.

7.7 How Do You Know You Have the Wrong Client?

There are two ways to identify that you have the wrong client. First is that they fail to provide you with those things you should expect in your internal consultant role. These are itemized in Section 7.4 of this chapter. When you fail to see your expectations fulfilled, you need to recognize that the initiative is in trouble; so too are your credibility and reputation for delivering quality.

The best thing to do in these circumstances is to talk with the client and reaffirm those things that they need to provide for initiative success. If this fails to be successful. then you need to determine how to extricate yourself and the team with whom you are working from the project. This will need to be done with your client, explaining to them why the effort is headed for failure. Because they do not want a failure of an effort which clearly identifies them as the client, they may alter their approach or decide to cancel the work. In either case, you have survived to work another day.

The second way you know that you have the wrong client is that, as the effort unfolds, it is clear that they will be unable to implement it. This takes us back to the second part of the definition of the client. Suppose that you begin working on a revision to the plant's work planning process with one of the planners. Although they may have good ideas about how to do this work, if they are not supported by the planning manager, there is little hope of implementation. You have the wrong client and need to change it to the planning manager or abandon the effort.

7.8 The Long-Term Client Relationship

Clients come and go. No client has a continual stream of work that they can assign to you. However, you need to keep in close contact with them so that the next time an initiative is identified, they will immediately think of you to act as the internal consultant for the effort. Staying in contact is easy. You can call or even stop in to visit. Remember face-to-face contact is the best way to stay in contact. But if you can't, then call the person on the phone and talk with them. Do not leave an e-mail or a voice mail. All too often these are deleted without the recipient ever hearing the message. You have a reason for the call besides making sure you stay in contact. Because you have worked for the client in the past, you need to call to see how the work related to the initiative is proceeding. Maintaining contact is a good approach; it is also good for your business. A good long-term client will create more work in strange ways. A dissatisfied client can undermine your ability to continue working and you may never even know why.

Five Things to Think About or Do

1. Make a list for yourself describing why you believe that you need clients to be able to successfully work as an internal consultant.
2. Think about an initiative that succeeded. Who was your client? What did they do or contribute to the process that helped it to succeed?
3. Think about an initiative that failed. If you had a client, was there anything that they did or failed to do that contributed to the failure? If you didn't have a client, can you see how the lack of this vital component had a role in the failure?
4. Make an internal consultant checklist so that you can explain to prospective clients what you will deliver as part of your internal consultant role. This will also help to clarify for the client what they should and should not expect. Use the checklist the next time you meet a new client.
5. Make a client checklist so that when you start working with new clients you can help them understand what they need to provide if the initiative is to be successful.

THE INTERNAL CONSULTANT'S WORK PROCESS

A good plan is essential to the internal consultant
It gets you where from here to there safely

8.1 Basic Internal Consulting

Imagine that you are contacted by a manger within your organization concerning a project for you to handle within your role as internal consultant. The project is very complex and involves many departments. Furthermore, it has a great deal of potential to improve your plant's competitive advantage in relationship to the other plants within your company. The project is to determine how to convert the highly-reactive maintenance organization into one that plans and schedules their work and is focused on equipment reliability vs rapid repair.

This is no small task. However, the manager is very enthusiastic and, based on input from other managers as well as from personal experience with your work, wants you to be the internal consultant on the project. The question we will answer in this chapter is how you go about the task of taking a manager's idea related to improved performance and convert it into a real, delivered work process for the organization.

There are specific steps in this process. If you have ever worked with external consultants, you will recognize this process because with minor variations it is applied by all consulting firms.

The steps of this process are:
1. The vision of the effort
2. Clarification of the assignment
3. Strategy development
4. Information gathering
5. Analysis and gap identification
6. Preparation of the recommendations
7. Presentation with defined next steps
8. Agreement—locking it down
9. Execution
10. Disengagement
11. Audit

There is also an ongoing process requirement that is part of most of these steps. That is validation of the work. Even though you believe that you clearly understand the manager's direction as defined in steps 1 and 2, you can never be sure without periodically validating that you are proceeding correctly.

At one point in my career, a manager asked me to review if our computerized maintenance management system (CMMS) could be used to track project costs. I proceeded to do the work per the steps outlined above. However, I never validated my direction so that I could be certain that the manager's expectations were being satisfied. At some point in the process, I got side-tracked because I had identified more problems with the cost control process—problems beyond the ability of the CMMS to track the expenditures. In reality, the whole process was broken.

This recognition and wanting to do something about it got me developing broader work process solutions. The work I was doing was far beyond what I was originally contracted to do. My presentation was well thought out; it had many valuable recommendations related to work process improvement. The problem was that what I had delivered was not what the manager had wanted me to do. I had failed to validate my direction periodically. As a result, I failed to deliver a good work product.

There is another aspect of the list that provides internal consultants with more involvement over a longer time horizon and with greater ability to impact change than their external counterparts have. This is step #11: the audit. For external consultants, their work often ends with the presentation and acceptance of the recommended next steps. Sometimes their work also enters into the area of executing the next steps. But ultimately, external consultants disengage. Not necessarily so with the internal consultants. You remain on site and often have the ability to watch the initiative unfold. This presence gives you an advantage over your external counterparts; it should always be included in the scope of your work effort.

The third advantage you have over the external consultants is time. Obviously they can not stay on site indefinitely and have to move on. What this means is that some of their proposed next steps, which require time for the site to accept and more time to develop and implement, may never get done. On the other hand, you are not going anywhere. You have the ability to file away until the proper time arrives those next steps that the site may not be ready to implement. At that time, you can then bring them out and with the "time being right" get them implemented.

When I completed a project designed to replace several separate computerized maintenance management systems with a standard one for all sites, I suggested as a next step to put a multi-site user group in place. The purpose was to ensure consistency across all sites. My idea was turned down. The time was not right. However, as an internal consultant I had the time to wait. Years later, I made the same proposal. This time it was accepted. The time was better and I was there to deliver what had previously been rejected.

8.2 The Process

1.   The Vision of the Effort

Before you are ever contacted by management to work on a project, they first need to have a vision of what they wish to achieve. Often this vision is supported by data, at various levels of specificity. At other times, it is not supported except by a subjective feeling that things are not as they should be. The vision, whether recognized as such, always comes first and sets the stage for all that follows.

Suppose that in our example the manager has worked for a prior company that performed their maintenance in a well-planned and scheduled manner. This knowledge provides an objective point of reference with which to compare the current highly-reactive situation. Therefore, the manager can be very clear as to a vision for the future. On the other hand, a manager who does not have this comparative information may have a less-defined vision, recognizing areas for improvement, but without the supporting data.

As an internal consultant, you often get called in on the initial stages of an effort where visions are or are not clearly defined. Your role at this point is very important because you need to work with the client to develop a clear vision of the future. This is essential work; without clear direction, you and the effort will flounder along the way. In that case, it is highly unlikely that the client will be satisfied with the end result.

There are many ways to clearly define a vision for the organization and for the effort in which you are involved. To this end, I would suggest reading my first book *Successfully Managing Change in Organizations: A Users Guide*, Chapter 4: *The Vision of the Future*. The information in this chapter should help you help your client to clarify the vision.

As an example let's assume that the vision is:
> Perform plant maintenance in a manner that involves detailed planning and scheduling processes in order to optimize the utilization of the workforce.

2. Clarification of the Assignment

Once the vision has been established, with or without your help, you are ready to move to Step #2. The fact that the manager has established a vision does nothing to clarify your role in the entire process. As an internal consultant, your role could run the gamut from facilitating the entire work process redesign to a more simplistic role of supporting an external consultant brought in to manage the entire process. Regardless what your role will be, you need to clarify it so that what you deliver as your end product is in line with the project champion's expectations.

There are many ways to go about clarifying your role. These approaches range from meeting to discuss the role you will play all the way up to developing and reaching agreement on a written contract that delineates the role you will play. I have been involved with all sorts of methods to achieve role clarification. The one that I have found to be the best is the written formal agreement of work scope between you and the client. This document accomplishes several things:

1. It documents the scope of work and clearly describes your role as a part of the scope.
2. It provides a document that can be referred to by both parties as the work progresses to assure that the effort is on track.
3. It can be used to control scope growth. This is a dangerous phenomenon that, once allowed to get out of hand, could cause even the most value-added effort to fail. The clarification document can allow you to reject additional scope and, hence, to control the work so that it stays in line with what was agreed.

Clarification documents often go beyond just the scope of work. They also can include things such as available funding, resource needs and how they will be satisfied, timing, and the latitude your client will be granting you as you work towards the project's completion.

In order to develop a document clarifying the scope of the effort, you need to meet with the principle clients, discuss and clarify the vision so that you understand it, and discuss both the scope and how your clients see your involvement in achieving it. Do not leave this meeting until you have these items clear in your mind. Taking detailed notes and not relying on your memory is highly advised. Once you understand all of these aspects of the effort, the next thing you need to do is to write the information down in a manner that you and your client will clearly understand. This may take several iterations until clarity is reached, but doing this is essential. Then, once finished, you and your client will be aligned.

For our example, let us assume that you will be the only consultant involved in the redesign and that issues other than scope are not a problem. Let us further assume that you will be facilitating the effort, which includes a detailed redesign of the work process to convert it from its current reactive state to one more closely aligned with the maintenance vision. Once this is clearly understood, written up, and agreed upon you are ready to move to Step #3.

3. Strategy Development

In this step, you need to determine the strategy which will be employed to deliver the expected scope of work. This strategy does not address the work outcomes. Instead, it is focused on how you will get to the point where you are ready to recommend action items to achieve the vision. The strategy really addresses how you will get the information necessary to determine how to proceed. The strategy also addresses how you will approach the information gathering process. Do you undertake the process alone? Do you build a team to handle this task and, if so, what type of team? Do you contract out the work to a third party who has recognized expertise in the area of information gathering?

Each of these strategies is viable and will work. However, certain strategies work better in specific situations than others. Table 8-1 describes four strategies and indicates when they are appropriate.

For our example, let us assume we are dealing with a single site where you are recognized as an internal consultant and have performed as a third party participating on several key change initiatives. Therefore, the strategy you chose to employ for the information gathering effort is to handle the work alone.

4. Information Gathering

Once you have determined the strategy for gathering the information, you need to go out and get it. The first question you need to address before you start your task is what information do you need to gather? Without clarity in this area, you may very well acquire the wrong information, leading you to erroneous conclusions and failure of the initiative.

Table 8-1 Information Gathering Strategies

Strategy	Why and When It Is Applied
You gather the information without support.	• You are considered by those on site as a trusted third party. • You have the ability to be objective about the information you acquire. • The information to be gathered is within your area of expertise. • The amount of information to be gathered or the number of people to be interviewed can be handled without negatively impacting the effort's timeline.
You gather the information as part of a site team.	• The effort is for a single site and the group has the ability to be objective about the information. • The volume of information to be gathered is more than you can handle in quantity or a time constraint has been imposed. • The client does not believe a third party is needed or a third party information gathering team is not available.
You gather the information with a company site team from other sites.	• The site personnel can not be objective about the information being gathered. • The volume of information is more than you can handle. • Multiple sites exist and have resources that can be provided or all sites are involved in the effort and you want a totally objective company view. • You want to provide exposure to people so they get a multi-site perspective.
You contract the work out to a third party.	• The volume of information to be gathered or the timeline constraints precludes a company team from doing the work. • Company resources are not available to be dedicated to this task, even for a short time. • The company representatives, regardless of the site they originate from, are entirely unable to be objective about the work effort.

The information that you want to gather needs to provide you with a clear picture of the current process and the areas where people believe there are issues that should be improved or corrected. There are several ways to go about acquiring this information. I have arranged them in order of importance from highest (best approach) to lowest (weakest approach). You can select any one you wish. But as you move down the list, the various approaches have a higher likelihood of providing you with less accurate information.

A. Individual Interviews

This is the best approach, although it is the most time consuming. In this approach, prior to the interview, you prepare a set of questions focused on the areas where you are trying to acquire information. You then individually meet with a comprehensive cross-section of people, seeking to gain the information you need about the current state of affairs.

If this process is handled correctly, you will discover that those you interview will have a great deal of information to provide. What you will also discover after the first half dozen interviews is that the information provided will be very similar. This is good because the interviews validate the information that was provided. As you keep hearing the same things over and over, you may get bored. Nevertheless, stick with it. The information is valuable and so is the feeling of involvement that you are creating in those being interviewed.

B. Individual Interviews Conducted by Two Members of the Team

This is an alternative to strategy A. The only difference is that you are bringing someone else from your team to be part of the process. By doing so, you are freed to ask the questions and have the dialogue with the interviewee while someone else takes the notes. It isn't a bad idea to interview this way; the exception is that some people will not open up and talk when there are two interviewers in the room. Another problem associated with this approach is that you are counting on someone else to take accurate notes the same way you would do if the notes were yours. This doesn't always happen, leaving you with missing information and frustration with the manner in which you conducted the interview process.

C. Individual Interviews and a Project Team Panel

This is an even further extension of strategy A. The problem is that interviewees certainly will feel overwhelmed if they have to sit in a room and be asked questions by a panel of people. I have seen this approach tried simply because the entire team wanted to ask and hear the answers to the questions. The problem is that the majority of people being interviewed in a panel situation will not provide the same information or detail as they would have provided in strategies A or B.

D. Group Interviews

This process can be successful if you wish to get a lot of general knowledge fast. In this scenario, you bring a group of people into a meeting and ask them questions in an attempt to generate a dialogue among the interviewees to gain additional information. The problem associated with this approach is that the quiet people will most likely never speak up. As a result, valuable information will be lost. Secondary problems with this approach are that it is hard to get a large number of people together due to their busy schedules, it is hard to control the process, and the results are not always of the value you seek. Nevertheless, you are often stuck with the information acquired because you probably will not be able to get the interviewees together again as a group or individually.

E. Interviews Via a Questionnaire

This is the weakest of all strategies because a) you have no direct contact with the people, b) you can not build a dialogue and, hence, explore other areas, and c) questionnaires are open to individual interpretation. If you ask a yes / no question, you will never get sufficient information to form objective opinions. If you ask questions where people have to write out their answers, you will get results that you will have to interpret. You may not interpret what the person meant correctly. In addition, there could easily be handwriting problems that could render the information unreadable.

In scenarios A–D, there are certain processes and rules that should be followed in order to get optimum results:

    a. Develop your questions ahead of time so that you can cover all of the areas where you seek information. There is nothing worse than an unprepared interviewer. It wastes both your time and the time of the interviewee. You will never be able to address all of the areas you wish to discuss nor get all of the information you seek.

b. Ask open-ended questions, not ones that can be answered by a yes or no. Open-ended questions leave room for the interviewee to expand, giving you more information related to the topic under discussion.

c. Select a neutral location for the interview. Having it in your office or the office of the person being interviewed is not as good as having it in a conference room or an unoccupied office. Neutral territory is always best.

d. Assure that you will not be interrupted. Interruptions break the chain of thought of both you and the person being inter viewed. This process is often difficult to restart after the interruption has occurred.

e. Assure confidentiality so that what is said doesn't get back to an individual's supervisor. You may find that, when you present your findings, a manager may ask you who provided you with the information. Never reveal your sources. For external consultants, it destroys their credibility during the engagement. As an internal consultant, it will destroy your credibility forever!

f. Feel free to ask clarifying questions to get additional information on a specific topic. However, as soon as possible, get back on track and continue on with your prepared questions.

No matter what strategy you choose, you will obtain information about the current process and state of the organization. However, there is more information that you need to acquire so that you can form a clear picture of the areas requiring improvement. This information lies within the documentation, procedures, and reports available to you as you dig up information about the current state of the process.

One revealing technique used to gain true insight into the existing work process is to physically follow the documentation through the actual process that handles it. In our example of trying to learn about maintenance planning and scheduling, you could literally attach yourself to a maintenance work order. Start out with production's identification of a plant problem. Then follow the work order through planning, scheduling, and work execution all of the way to work completion. If you follow this process in every detail, talk to all of those involved to really understand what they do and how they do it. You will gain tremendous insight and a great deal of pertinent information. This, coupled with the information you obtained from the interview process, will position you for the next step.

5. Analysis and Gap Identification

Once you have completed your interviews and the other parts of the information gathering effort, it is time to conduct the analysis and identify the gaps. If you performed Step #1, identifying the client vision, you should have a good idea of what the client wants to accomplish. At this point, you also have a clear idea of where the organization stands related to the vision. In some cases, the gap between vision and the current reality may be small and, in other cases, it may be large. It is unlikely that your services would have been retained if the former was the case.

Your task at this juncture is to take all of the information you have gathered, identify the gaps, and make recommendations for gap closure by the client.

There are many ways to do this analysis. However, the key thing to remember is that it is your job in this phase to develop a clear picture of the current state along with factual information to support your conclusions. If you have interviewed alone, then this task falls solely to you. If you have interviewed with a team, then be prepared for long team meetings as you sort through the information and try to gain a common understanding of the site's current situation.

There are three aspects of the analysis phase that need to be considered: the hard skills, the soft skills, and the organizations culture. Each of these will play a role in what is actually going on at the site. Hard skills are the actual tasks that are being performed. Soft skills are the elements that support successful performance of the hard skills. These include leadership, work process, structure, group learning, technology, communication, interrelationships, and rewards. The organizational culture—which includes organizational values, rites and rituals, role models, and the cultural infrastructure—consists of the essential controlling factors for all current tasks as well as those which you may wish to implement or alter. For this reason, they need to be taken into account as you develop your recommendations. There is further detail about soft skills and the organizational culture in Chapter 5 and even greater detail in my book Improving Maintenance and Reliability Through Cultural Change. If you wish further clarity on these very critical elements of change, I suggest you read this material.

In the analysis phase, you need to understand clearly what is happening currently in the site as it relates to your task, which is to develop recommendations to close the gap between the current (as is) state and the vision (to be) state. The question is: How can you effectively and efficiently handle this task?

I have a format I have followed that ties in the hard skills, the soft skills, and the elements of culture in a way that will let you put all that you have learned in the information gathering phase into perspective.

Take a blank piece of paper. Across the top of the page, write the vision statement for the effort so that you will keep it in focus. Draw three vertical lines to provide you with three columns. Label them Hard Skills, Soft Skills, and Organizational Culture. In the columns for soft skills and culture, list the individual elements that are the component parts of each. This is shown in Table 8-2.

Now ask yourself, "In my interviews and other information gathering efforts, what did I learn or discover about the current (as is) state at the site and their effort to achieve the vision?" List the information in the proper place on the page by associating it with the various sub-elements. On a separate piece of paper with the same format, list those things you learned that you believe would block the change. This is good reference material.

In our example, the client wanted to move away from reactive maintenance and institute a planning and scheduling process. Table 8-3 summarizes some of the information gathered during the interview phase of the effort. It is shown in the same format described in Table 8-2, which you should use for this effort. You will also notice that this information provides you with insight into conditions that will serve to block the change.

Client Vision:		
Hard Skills	**Soft Skills**	**Organizational Culture**
	Leadership	Organizational Values
	Work Process	Role Models
	Structure	Rites and Rituals
	Group Learning	Cultural Infrastructure
	Technology	
	Communication	
	Interrelationships	
	Rewards	

Table 8-2 The Analysis Form

Client Vision: Perform maintenance, utilizing planning and scheduling to improve workforce effectiveness and efficiency		
Hard Skills	**Soft Skills**	**Organizational Culture**
A planning process was tried but failed due to lack of support by both Production and Maintenance. The current managers in these departments were the ones involved. In some areas, planners exist but they do no real planning of the work; they provide daily field support. Pre-scheduled work seldom gets done per the schedule because the work crews are always moved to the "perceived" problem-of-the-day.	**Leadership** They claim to want planning and scheduling but have not supported the effort.	**Organizational Values** The values of the site are reactive and focused on the day-to-day problems of keeping an unreliable plant operating.
	Work Process A planning and scheduling work process does not exist at present. The work is scheduled based on the immediate needs of Production.	**Role Models** The role models are all reactive responders. They are praised by their customers and receive the promotions based on this behavior.
	Structure There is no planning and scheduling jobs listed on the organization chart. Planners exist in some areas but there is no supervisor for this effort; they report to execution. There is no defined scheduling function.	**Rites and Rituals** The rituals (work tasks) are all focused on reactive response to the problem-of-the-day.
	Group Learning While the site recognizes the value of planning and scheduling, they still believe their reactive approach works well. There is documented data to show the lack of efficiency of the maintenance team.	**Cultural Infrastructure** Keepers of the Faith (mentors) are focused on reactive work response. Story tellers focus on how key people responded to problems and save the day. Gossips and spies undermined prior efforts to change.
	Technology The CMMS can support the planning and scheduling function, but it is not used.	
	Communication The work assignments are verbal from production to the maintenance foreman. This often results in serious confusion.	
	Interrelationships Strong between Production and the maintenance foreman. Poor between higher levels of management.	
	Rewards The rewards are based on successful reactive response. There is no reward nor recognition for a well-planned job.	

Table 8-3 The Planning and Scheduling Analysis Example

You may notice that some of the information may fall into more than one category. This is all right because even though a piece of information is identified more than once, the different categories will provide you with different perspectives.

Once you have completed sorting out your information using the process described, you should have a very clear idea of the problem areas that need to be addressed in Step #6.

6. Preparation of the Recommendations

Although you and your client may have a good idea of what needs to be done to achieve the vision, how recommendations from your analysis are prepared and presented is critical.

I have found that there is a good way to develop your recommendations for presentation to management. PowerPoint is a very good tool for this purpose. It allows you to present the ideas and, if set up correctly, leaves a great deal of room for discussion. There is one thing of great importance to a successful presentation. You should not use any slide with words smaller than 24 font. First, you don't want to present the full report on your slides—just the key ideas. Second, words displayed in fonts smaller that 24 are difficult, if not impossible, to read. PowerPoint presentations that the audience can't read will cause a loss of attention as they try to read the material. Needless to say, as they are doing this, they are not listening to what you have to offer.

A good format for these presentations, one that eases people into the recommendations for change, is structured as follows:
   a. The cover page.
   b. Vision statement from the client so that it is not perceived as yours.
   c. Methodology employed in gathering the information.
   d. Major findings from the information gathering effort. List each finding separately and provide supporting bullet points.
   e. Minor findings which include items of interest and things that are not a major impact.
   f. A high-level list of next steps for the effort.
   g. Suggested detailed next steps to keep the effort moving. Remember the presentation is the end of the beginning of

the effort, not the effort itself.

h. Question and answers to enable you to take questions from the audience and clarify any points you still want to make.

The slides are the heart of your presentation. You should also prepare a formal written report describing the same material as you had in your presentation. Don't make it too long or it will never be read. On the other hand, making it too short will not provide sufficient detail to make the material valid in the view of the reader. I have found that a good solution to making the report a proper length is to provide the details for each slide in the slide's notes section. In this way, you can print the slides with the notes as your formal report. It helps to keep your audience focused because they can relate the slides they saw in the presentation with the report.

Another important point is to keep the presentation and the report generic. Avoid any reference to any person or group. Using the third person in reference to the site helps to keep the presentation at a factual level at all times. It also serves to remove defensiveness and encourages discussion.

Suppose you want to state a finding and make a recommendation about reactive maintenance.

How not to say it

This information, although valid, attacks the client and their organization.

Finding 1: Those interviewed conduct maintenance in a poor and highly-reactive manner. Planning does not exist and the planners are used to support field execution.

Recommendation: Management needs to implement a planning and scheduling process; it must stop performing maintenance activities in the current inefficient manner. Planners must be used for planning purposes.

A better way to present the finding and recommendation

Finding 1: The work process at the site appears to be highly reactive. Although there are identified planners within the organization, they do not appear to be effectively utilized.

Recommendation: In line with the vision, work processes that effectively utilize the planners and deliver an efficient planning and scheduling process should be implemented.

If you carefully examine both statements you will recognize the generic nature of the second way to present the information. In a generic format, you are not attacking anyone. People will be more willing to listen and hopefully support the steps to implement it.

7. Presentation with Defined Next Steps

Once you have sorted through all of the information you gathered and developed your presentation, it is time to present it. If you have followed my recommendation about validating your work with the client, then there should be no surprises. I recommend you review your presentation with the client first; a sort of "dry run." In that way, any minor adjustments that are required can be made before the actual presentation is made.

I also suggest that once the client's initial review is completed, they be allowed to determine the audience for the presentation of the findings and the recommendations. Having made numerous presentations of this sort, I have found that each client has a different approach. Some want only their staff to hear the presentation first, with later presentations to a larger and often more mixed audience. Others want exactly the opposite. Of course, you can recommend the audience; to this end, I recommend that you start with the client's staff. Through successive meetings with larger audiences, the information can be disseminated. This approach gives the leadership team a chance to digest the material and buy into the steps to make the change before the information is presented to others.

There is also another reason for meeting with the client before the presentation. It gives them the background information they need to support the next steps. At the actual presentation, a preview meeting with the client enables them to show an understanding of the material as it is presented. It lets them develop some leading questions that will help you and the organization move forward with the recommended next steps. They can show their involvement and use the meeting forum to get others involved.

Things to remember for your presentation include:

a. Make copies of the presentation for the group using the PowerPoint notes format. However, do not give out the material until the end of the presentation. If you give it out beforehand, people will be reading it and not paying attention to what you have to say.

b. Use a projector to show your slide presentation. Often people will hand out the slide presentation and talk the group through it without projecting it. This is very ineffective.

c. Stand up when you present the material. In this way people will be forced to focus on you and hopefully on what you have to say.

d. Stay on track both for the content and the time. You will have so much information that it would be easy to spend hours discussing it. Stay focused or risk losing the group.

e. Don't get defensive if challenged. State your facts. If some one disagrees, ask the audience whether they feel the information is valid. If they won't acknowledge the facts you are presenting, then they most likely will not support the associated change.

f. Talk to the group, but focus on the client. If the client (usually the senior manager) acknowledges your findings, the rest of the group probably will too, at least overtly (see Chapter 9 on resistance).

g. Based on the structure of the findings from Figure 8-2, you may have to teach soft skills and the organizational culture. Very few maintenance and reliability people are familiar with these concepts because their normal focus is on hard skills.

h. Set up the presentation for mid-morning. If you serve any refreshments, don't serve coffee cake or anything with sugar. The mid-morning time slot works well. People have had a chance to get their daily activities started and they are still fresh. Luncheon meetings are distracting and the after noon meetings are when people get tired or when they are thinking about the end of the day issues and deadlines.

i. The issue of food is also important. Cakes and other snacks with high sugar content gives people a sugar rush, but as this wears off, they become sluggish. If you feed people snacks with high sugar content, you risk losing your audi ence at the end of the presentation when you need their participation the most. Typically in presentations of this sort the most I provide is coffee and tea.

j. Have someone you trust act as the scribe and take notes during the presentation. Having someone else perform this task frees you up to make the presentation and lead the discussion of the next steps. Doing both well is virtually impossible.

The last thing about presenting your findings and recommendations is to make certain that you discuss and end the meeting with defined next steps, along with named responsibilities and completion dates. If possible, don't end the meeting without accomplishing these tasks. You may not get an opportunity to assemble the group focused on change-related initiatives again. Without next steps, you will delay the effort; at worst you may kill it.

8. Agreement—Locking It Down

Following your presentation and agreement on next steps, you need to publish the presentation minutes and agreements. Don't wait! Do this work within one day of the meeting; otherwise, people will forget what they agreed to do. Also use these notes as a validation point with your client. What you want to validate is the client's understanding of the agreements that were made, responsibilities, the timeline for accomplishing the work, and your role in the overall effort. At this point, you have reached a major milestone and are ready to move into the execution phase.

9. Execution

The execution phase of the work begins when you complete your presentation and reach agreement on the next steps, as well as when you finalize your role as the internal consultant on the upcoming effort. Execution ends partially with deployment.

However, because we are dealing with a process change, there really is no end for the site personnel.

For you, the internal consulting part of the work effort ends with your disengagement. There is also an audit phase in which you may become involved. This is one advantage internal consultants have over their external counterparts. They remain on site even after they disengage; they can get re-engaged to help with the audit of the process after it has been in effect for a pre-determined period of time. There is also the possibility that you will have enabled the site personnel so that they will do the audit themselves and only call on you for advice.

Working through the work redesign process will be covered in Chapter 11 on teams. There are two other critical parts of the work: readiness for change and sustainability. These will be covered in Chapter 15.

10. Disengagement

As an internal consultant, you will have the opportunity to work on many site-wide or even company-wide initiatives. One thing that we discussed early in this book was the need for site ownership of the initiative throughout its entire life cycle. If you own it, failure of the initiative will be the ultimate outcome. In addition, you might as well sign on as a part of the organization because you will never be able to leave.

A disengagement strategy should begin as soon as you are assigned to work as an internal consultant on the initiative. At the outset, you need to make sure that there is an owner, a site initiative lead, and a project team. The last of these, the team, includes you filling some, if not all, of the internal consultant roles described in Chapter 10. The owner or lead role should never be yours. Otherwise, the ownership you need to disengage will be lacking.

As the effort proceeds, you need to make sure that you do not ever become the owner or leader. This is a self-monitoring process that works on two levels.

a. You need to make sure that the designated leads do not abdicate their responsibilities to you. Many initiative leaders will try because they have "more important" things to do; they see you as someone to whom they can delegate tasks.

b. You also need to make sure that you don't try to take over. By the very nature of their jobs, internal consultant types are self-motivated and "can do" type people. As a result, if they see a leader floundering, they will jump in and take over. In many cases, the leader may allow it. Site ownership is then transferred and the ability to disengage is lost.

Assuming that you have set up the parameters of disengagement, there should not be much of a problem turning the last few aspects of the work over to the site. This should be a formal process with an established turnover date so that what is taking place—your exit—is clear to all.

Signs to look for that indicate that your exit is causing a problem are as follows:
a. Signs of fear and disorganization in the team
b. The effort stalls without your involvement
c. Continuous calls and e-mails for help
d. Calls from the client with expressed concerns
e. Action item dates not being met
f. People reverting to the old way of working
g. Personal feelings of discomfort when you talk to people about the initiative and the progress it is making
h. Other forms of resistance (see Chapter 9)

If many of these signs are prevalent, then you have not successfully disengaged and you need to find out why and take the necessary corrective action. This might even mean re-engaging to determine the root causes of the problem and fixing them.

11. Audit
Change is a continuous process. It is not a project that you and, more specifically, the site team can complete and then walk away. For this reason it needs to be monitored. Problems should be identified as they appear and corrective action taken to bring them back into alignment with the overall goals.

If they truly have ownership of the work, then this is a site effort. However, as an internal consultant you can support it or, if necessary, even lead the effort. It is almost the same as the information gathering process. The difference now is that you are identifying gaps between the vision and the current state of the work initiative, not the former "as is" work process. Periodic audits are extremely important because they alert you about process problems which can then be addressed before they gat to the point that they can not be corrected.

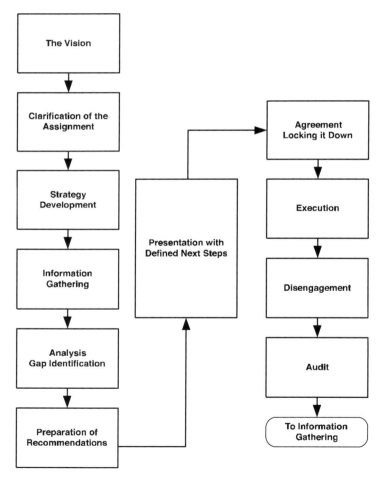

Figure 8-1 The Internal Consultant Work Process

The eleven steps described in this chapter can be represented by the diagram shown in Figure 8-1. The series of blocks on the left are those steps required to gather the information, make the analysis, and prepare you for the presentation of findings. The blocks on the right are the steps that follow the presentation. I have shown the presentation block in the center because it is the central task of the entire process.

As you can see, the internal consultant work process can be rather complex and time consuming if it is done correctly. It is clearly not a part-time job for anyone. Consequently, there is a need for experienced, well-trained internal consultants. If this is not possible, then external consultants must be hired to handle the work correctly.

Five Things to Think About or Do

1. List the eleven steps of the work process and identify which ones you have used in the work that you have done. Why did you not use the others?
2. Identify which steps worked well and which ones caused you problems. Try to identify why the problems existed and what you might do differently next time.
3. How have you clarified internal consultant work assignments and your role in these assignments in the past? How would you go about it in the future? Prepare an outline of how you would approach this task.
4. Review the presentation task. How have you done presentations in the past? Would you do anything differently the next time you had to make a presentation? What would it be?
5. Think about an initiative in which you were involved that did not succeed. Was it audited before it was declared a failure? If it was audited, what was discovered and what was done to try to keep the effort from failing? If it wasn't audited, would an audit have helped save it from failure?

RESISTANCE

That which doesn't kill you will make you stronger

9.1 Resistance Discovered

Suppose that one of your initiatives as an internal consultant was to work with the site personnel to develop a detailed planning and scheduling process. The process was to focus on detailed work plans and deciding a week in advance the planned jobs that would be executed in the coming week. As part of the work process, you held meetings that included not only Maintenance, but also representatives from your customer, Production. This work took your team a considerable amount of time in order to develop all of the details of the process and to make certain that the requirements of all those involved were addressed. As the process was completed, you and your team launched into the training phase; over several weeks, you trained all of the maintenance and production personnel.

Then the much anticipated day of deployment arrived. At first everything appeared to be functioning well. However, after several weeks you noticed that the number of unplanned jobs was still at the same level as before the new process was deployed. You know this couldn't be correct because the new process was designed to eliminate virtually all of the unplanned work. As you investigated the problem, you discovered that the production supervisors and their maintenance counterparts simply decided not to follow the process. Their logic was that the work the crews were doing needed to be done now; it couldn't wait several weeks to go through the planning and scheduling process. Essentially they were telling you and the work process team that they were not going to do what was asked.

In another example, suppose that your plant deployed a new preventative maintenance program. This new program required the mechanics to work in a much different manner than they had in the past. Previously, they were assigned to the production line and performed the work assigned to them by the line operators. When these assigned tasks were completed, they returned to their staging area and awaited a call with their next assignment. In the new program, they were being assigned daily preventive maintenance tasks that had to be completed to keep the program in compliance. In addition, they were reassigned from working for Production to working for a preventive maintenance foreman whose job it was to keep the program moving.

As an internal consultant, you were given the task to discover why the program never was able to stay in compliance. You also had to determine why, even with documented preventive maintenance being performed, the equipment was still failing at the same rate as before the program was introduced. After some investigation, you discovered that, although the mechanics were performing the preventive maintenance tasks, they were not doing them on a timely basis nor were they doing them very well. There lay the lack of improvement in equipment reliability.

In another example, suppose that you and the team have determined that having line operators do minor maintenance will help improve equipment reliability; they will make minor repairs before the problems become major. It also will help the productivity of the maintenance department by having the minor repair tasks handled by the operators, leaving the maintenance crews to work on the large, more complex ones. After the team has developed the scope of work, purchased and distributed the tools, and trained the operators in how to perform minor maintenance tasks, the work initiative is deployed. Within days, it grinds to a halt because the tools that were distributed are missing. To keep the effort moving, the project team purchases additional tools. Once again they disappear and the work initiative falls apart. Through sabotage of the process, the operations have caused it to fail before it is ever allowed to get off of the ground.

In one more example, suppose you and your team develop a process to improve materialization of the planned work. During the training sessions, everyone agrees that this new process will vastly improve how the maintenance jobs are materialized. However, upon deployment something is obviously wrong. Although everyone appears to be going through the motions associated with the new process, the foremen and planners are still making the same number of trips to the storehouse in order to obtain job-related materials. On the surface, the process appears to be followed, but it is obvious that this is not actually the case. It is not being followed. Nothing has changed in the approach towards obtaining material for the maintenance jobs.

These four scenarios bring up a very interesting question. Why do change initiatives developed by intelligent and caring individuals not always succeed? You may start with a good idea. But somewhere between the creation of the good idea and its actual implementation, the process gets interrupted. The change, as good as it is, fails. In this chapter, we look at why this happens and what you can do to keep change moving forward.

9.2 Resistance as a Part of Change

Change often fails due to bad planning or bad execution. But what if you have taken the time to carefully plan the change effort and have worked hard to execute it? The process you put in place still can flounder. It may never reach its full value or, worse, fails entirely. The critical factor may be resistance to change, a condition present in every change effort and in every work environment. In spite of your good ideas, many people will see them in a different light. They believe that the change is not in their own best interest or that of the company's. They then act on this belief, doing whatever they can to stop the change from happening.

Some people who are involved in developing change initiatives have the attitude that their position of power, or even the simple obvious value of their idea, will outweigh those resisting; therefore, in the end, the change will take place. This opinion is often held by site management or even the consultants who are hired to help develop and deploy the initiative.

Yet this approach is extremely dangerous. Resistance, if ignored, can potentially destroy even the best idea. At the least, it will severely undermine the effort and its potential value. If, however, you understand what is going on, you can take preemptive action to address it as part of the process design and, thereby, minimize the problems that unaddressed resistance can provide.

9.3 Why Do People Resist?

Why do people resist change? The simple answer is that they do not view the change as an improvement, even if you and your team do. Often, if asked, people will tell you they see it as a step in the wrong direction and not in either their best interest or the best interest of the company.

Sometimes they are correct. But often their feelings are clouded by their emotional response to a perceived mismatch between the new environment (what you are trying to do) and their comfort zone (the area in which they operate on a daily basis and in which they feel a certain degree of comfort). Within this comfort zone, they do not feel threatened by either the work or the environment. This state is often called the status quo.

Take people out of their comfort zone and they not only feel uncomfortable. They also do whatever they can to restabilize their environment. Sometimes this is easy for them to do. But at other times, protecting their environment can be difficult or outright impossible. At these times, an individual's level of stress increases; they try even harder to restore the status quo. Thus, a critical component of this comfort zone model is that the further you take people beyond this zone, the more that stress levels increase to the point where they become unbearable. The stress can become so severe that people try to restore the status quo by resisting the change.

Consider the four examples at the beginning of this chapter. In the first, the planning and scheduling initiative failed because the operators and maintenance foremen involved didn't like the new process and simply refused to do it. In the second example, the mechanics did not like the idea of a structured preventive maintenance program directed by a foreman vs their existing process directed by the operators. Furthermore, they did not like having to perform scheduled preventive maintenance. They found much more enjoyment in doing the tasks that they were asked to perform in order to keep the production line running. Unlike the first example, they did not feel that they could simply refuse to take on the new job tasks. Their experience was one of punishment accompanying refusal to do work. Nevertheless, they still resisted by doing the work poorly in hopes that management would become dissatisfied with the initiative and allow them to return to their status quo.

In our third example, which focused on operators doing minor maintenance, the operators also did not feel that they could simply refuse. Yet they still had a strong desire to resist this new initiative. Their solution was simply to get rid of the tools needed to do the work. After all, without the correct tools, they could not perform minor maintenance and could return to their status quo. The fourth example, work materialization, was also resisted by those given the materialization assignment. Their resistance took the form of a low-key refusal. Instead of outright refusal to do the work, they went through the motions so that those looking in from the outside would think the work was being done when, in actuality, it was not.

9.4 Forms of Resistance

As we have seen, resistance takes on many forms. You need to recognize them before you can properly address resistance and have a successful work process change. Some forms of resistance are obvious and easy to recognize. Others are very subtle; if you are not paying attention, they will undermine or even destroy the change effort before you can react.

Resistance can be categorized in four forms, depending whether the resistance is active or passive and whether it is open or hidden. Figure 9-1 illustrates these four forms.

The y-axis measures the visibility of the resistance by those affected by the change. This action can be open; the resistance to the new process is obvious for all to see. The action can also be hid-

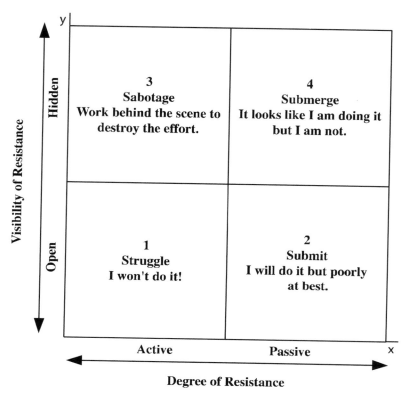

Figure 9-1 The Four Forms of Resistance

den. In this case, the resistance is below the surface, not easily seen. If you are paying close enough attention, you may notice it, but it is more difficult to identify.

The x-axis identifies the degree of resistance, whether it is active or passive. Active resistance is designed to stop or hamper the change process. It can be destructive to the organization because, as the resistance evolves, people may be forced to take sides, leading to substantial business problems. Passive resistance is more difficult to confront because it is generally less obvious. Rather than trying to actively block change, people engaged in passive resistance are not going along with the effort. They might continue to use the old system, ignoring change for as long as they can. They may work more slowly, take extra time off, and keep "forgetting" how to incorporate the change into the work process.

As Figure 9-1 illustrates, the resistance model has four quadrants. Each is discussed below.

Quadrant 1—Active, Open Resistance—Struggle ("I will not do it.")

You begin to implement change when some of the employees openly tell you the change is wrong or, even worse, they won't go along with it. A scenario like this one should never happen. Change should be discussed and buy-in achieved before you try to implement it. If you have planned the process change and the subsequent implementation, your organization should never have to be placed in this situation. Otherwise, you can expect active and open resistance. The one good aspect of open resistance is that it indicates that your employees feel comfortable enough with you as the manager to tell you openly how they feel about what you are doing. If they don't trust you, the resistance will be hidden, leading to Quadrant 3 behavior.

Quadrant 2—Passive, Open Resistance—Submit ("I'll do it but poorly, hoping you will make it stop.")

Passive resistance is quite different from active resistance. It involves people submitting to the new order of things, in a sense "going along." Don't mistake this submission for acceptance and open embrace of the change. Even though they do what's necessary to make the change work, you lose their energy, enthusiasm, and loyalty. Unless you win them back, you may see a gradual downturn in productivity and increased turnover.

Quadrant 3—Active, Hidden Resistance—Sabotage ("Work behind the scenes to destroy the effort.")

If your employees do not trust you, resistance to change will take on a different form. In Quadrant 1, employees didn't feel they were threatening their own security if they told you exactly how they felt about the changes you were making. In Quadrant 3, they do feel threatened. They will resist change as actively as the employees in Quadrant 1, but will try to hide their resistance, sabotaging your efforts. At least in Quadrant 1, you knew what resistance you faced and could respond to it. Not here. The problem often comes from management style. Employees want to be trusted. If, however, the management style at your company is that employees should simply do what they are told, then resistance will be hidden and you will face sabotage.

Quadrant 4—Passive, Hidden Resistance—Submerge
("It looks like I am doing it, but I'm not.")

Unlike Quadrant 2 resistance, Quadrant 4 resistance is hidden. Because resistance is passive, it frequently is not as severe as sabotage (Quadrant 3). Nevertheless, it is still dangerous. Your employees are indirectly saying that they will do what's asked, but will undermine the effort at every opportunity. At least with active, hidden resistance you are aware of the resistance once it has occurred. With Quadrant 4, you can not see it; it is submerged. On the surface, everything may seem fine. Meanwhile, below the surface, you face severe problems. Quadrant 4 resistance is used to undermine change. Your process may fail and you may never know why. Nor will you have anyone specific to blame for the failure, except maybe yourself.

9.5 Coping with Resistance

Once you know about resistance and can recognize it, the question becomes, "How can you address it and still have your initiative be successful?" Sometimes you simply can not, but these instances are in the minority. Most of the time when resistance is recognized, it can be addressed as part of the initiative. The fears and emotions that are causing it can then be accounted for in the development.

All too often, the people who are part of the project team, as well as the internal consultants who are supporting the effort, view resistance as something that needs to be overcome. They view the resistance as a force threatening to ruin the work initiative, and those who are resisting as the enemy. Quite the contrary, resistance needs to be addressed head on. The issues that are causing the resistance, as well as what the people are feeling regarding the change, need to be addressed. Viewing resistance in whatever form it takes as an enemy of change is not the correct approach.

Consider the quote from Machiavelli stating that those

involved in the change have either "enemies in those who did well under the old process and at best lukewarm defenders in those who feel they may do well under the new." These people are the resistors in whatever form they may take. Wouldn't it be better to win over the enemies and convert the lukewarm defenders into staunch supporters? Addressing resistance as part of the change effort can make this happen.

Let's look from a business perspective at how resistance emerges. People generally resist when they move out of their comfort zone. The basic cause is often the lack of agreement between the goals that you are trying to achieve and the goals of the individual or group with whom you are trying to achieve them. If the goals, or change initiatives, are in agreement with the employees' goals, you have no problem; the comfort zone is not being violated. The comfort zone is expanding with the agreement of the employees who have to adjust to its growth.

The problems begin when your goals are not aligned with the employees' goals. The key then is to put change initiatives into place in a way that allows the goals of the two forces (you and the employees) to be aligned. With alignment comes success. Your success is shaped by two important factors. The first factor measures how well the goals agree between those trying to implement them and those having to do the work. The second factor measures the balance of power between the groups. These factors dictate the approach you need to implement the change process successfully.

Figure 9-2 provides a four quadrant diagram that illustrates different ways change is implemented. In this figure, the x-axis measures how well your goals are in agreement with those of the employees who have to implement the change. The y-axis measures from low to high the balance of power between those who are trying to implement a change initiative and those who are responsible to make it work.

The balance of power, in conjunction with the amount of agreement on goals, determines the level of resistance to the change initiative. A high balance of power corresponds to a team relationship. Both sides work together to accomplish the change efforts. A low balance of power indicates that you and the management team will dictate to the employees what will and what will not be done. This relationship could easily be described as "com-

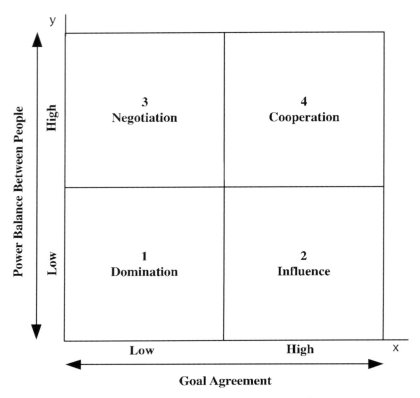

Figure 9-2 Strategies for Implementing Change

mand and control." In addition, the balance of power influences the way in which the change initiative is presented to the organization. It affects the type of resistance that may emerge. The balance of power can also help you determine the best way to rollout the initiative so that you minimize conflicting goals and resistance.

Quadrant 1—Low Goal Agreement, Low Power Balance— Domination.

In Quadrant 1, you and the employees share no agreement about the goals of the organization. In addition, the power balance is low; when goals are presented, they are usually in the form of "here is what you are going to do." The combination of low agreement and low power balance creates an environment in which you dominate the employees.

In a domination model, resistance usually takes the form of sabotage (if employees resist actively) or submergence (if employees resist passively). You tend to see submergence more often. Those who openly resist generally no longer work for the company, whether by their choice or by management's. The evidence of submergence is that, after dictating the goals, management repeatedly wonders why change is never implemented or, if it is, it is superficial at best. When the organization acts in this manner, it usually doesn't recognize that its entire approach to the process of change is wrong, let alone understand why its initiatives are failing. The best way to avoid resistance in this quadrant is simply not to be in this quadrant. In today's business environment, the best managers engage the employees, rather than lose all of the value that they could add.

Quadrant 2—High Goal Agreement, Low Power Balance— Influence.

In Quadrant 2, you and the employees agree on the goals; you are in alignment. With this relationship as a staring point, change is much easier to implement because there is no conflict about what needs to be accomplished. The only downside is that the power balance is low. This imbalance doesn't usually lead to a great amount of resistance, but you still may encounter some.

Because the power balance is low, you must use influence to get your initiatives implemented. Although the employees agree with the goals, many problems can occur, based on how you apply influence. If you take advantage of the power imbalance and try to dominate the employees, even those who agree with the goals will resist your efforts. Influence can be applied in many ways: personal charisma, rewards, and cheerleading among them. Especially because the power is on your side, employees will be encouraged if you take time to listen to their views. You need to find what works best for you. One easy method is to engage those who you are trying to influence. Gather a group of key personnel, discuss what you are trying to achieve, and use their collective input to determine how to best influence the work group.

Quadrant 3—Low Goal Agreement. High Power Balance—Negotiation.

In Quadrant 3, the balance of power is shared. Therefore, you and the employees are in a good position to work together to implement the change effort. Unfortunately, the agreement on the goals is not strong, leading to the potential for resistance. Your best strategy, based on high balance of power is of negotiation. In many ways, your strategy here is similar to your Quadrant 2 strategy. However, the methods of reaching goal agreement are more sophisticated. In fact, it is through the negotiation process that goal agreement will be reached.

How you manage the negotiation process will determine the kind and degree of resistance. As you may already know, there are many types of negotiations. The traditional model still used in too many union-management negotiation processes is the win-lose model. Each side puts its issues on the table. Then through a process of withdrawing or conceding items, the two sides eventually (we hope) arrive at something acceptable to all. To get to this point of agreement, much is often lost. What the two parties must concede to reach consensus influences their level of resistance. Often, this resistance causes this approach to break down. The better approach is to conduct negotiations in a manner that leads to win-win result in which both sides are happy. Many books specifically about negotiating are available on the market. When you conduct a win-win negotiation, the focus should be on what's good for the business. Conducting a win-win negotiation is not easy. You may want to have an experienced facilitator present.

Quadrant 4—High Goal Agreement, High Power Balance—Cooperation.

Quadrant 4 is relatively easy. In this quadrant, you and the employees share a balance of power and agree on the goals. The change initiative should be successful. Furthermore, you should encounter no resistance to implementing it.

9.6 Working to Address Resistance

In each of the quadrants of Figure 9-2, there is an implied statement that the resistance will be addressed and the goals of the project team and the organization will fall back into alignment. Alignment may be achieved based on the concept of managerial power balance, where the team dominates or influences the organization. Or it may be one where the power balance is at a peer level and resolution is reached through negotiation or cooperation. In either case, the agreement and alignment is implied.

The issue you need to answer is how you, as an internal consultant, can get these groups into alignment while at the same time addressing the resistance? The answer is either to show how the change initiative delivers value far in excess of the current situation or to identify the forces that support the sustaining of the status quo, then diminish them.

Getting groups back into alignment, addressing resistance and helping the initiative to move forward can be accomplished through a process called Force Field Analysis. Other methods work equally well, but I have found Force Field Analysis to be the best

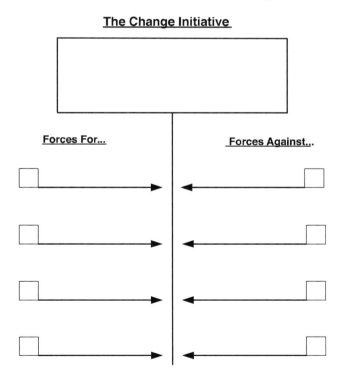

Figure 9-3 The Force Field Analysis Diagram

and most direct process. The idea behind this process is that you identify the forces that would make you want to initiate the change as well as the resistance-like forces that would make you not want to implement the change. One all of these forces are identified, they need to be ranked by level of importance. The site can develop their own scale, but I like a scale where 1 is the lowest level of importance and 5 is the highest.

Once the priority levels are set, the goal is to determine how to lower the resistance scores and / or elevate the forces driving the change. Then everyone can see how the change will be the correct way to proceed delivering the highest benefit. At the same time this process will hopefully indicate to those resisting the change that there is value in their acceptance.

Figure 9-3 shows a force field diagram. The change initiative is presented in the block in the center of the diagram. The forces for the change are listed on the left and the forces against on the right. As we have indicated, each is scored from 1 (lowest importance) to 5 (highest importance). The importance rankings for each item is entered in the small box next to the identified item.

To clarify how this process works, let's look at an example. Suppose that in your plant there are maintenance mechanics assigned directly to the production line. These mechanics are waiting for Production to identify maintenance problems. It is then their task to go and make the repairs. The plant has implemented this process in order to affect timely repairs to the equipment. They believe that making repairs in this manner will keep them from growing from small problems into big production-interrupting ones. Initially this process was very effective, but over the years it has degraded. To determine how to make more effective use of the current mechanics assigned to the line, a project team was assembled. One of the plant's internal consultants was brought in to facilitate the effort.

After a great deal of analysis, the team determined that there were significant advantages to moving the mechanics back into the day-to-day maintenance work crews and teaching the line operators minor maintenance tasks. As the deployment process was initiated, a great deal of resistance was encountered from Production. They believed that they were losing a valuable resource. The problem was becoming serious and it was necessary to address the resistance before implementation.

To accomplish this task, the internal consultant turned to Force Field Analysis. Maintenance, Production, and the project team met; they developed a list of forces for the change along with importance ratings. They also developed a list of forces against the change with a similar set of ratings. This force field diagram is shown in Figure 9-4.

The next step of the process involved explaining the "forces for change" in detail so that those resisting the change understood what was trying to be accomplished. This had some impact in that it explained the change based on factual information obtained during the data gathering process of the internal consultant's efforts.

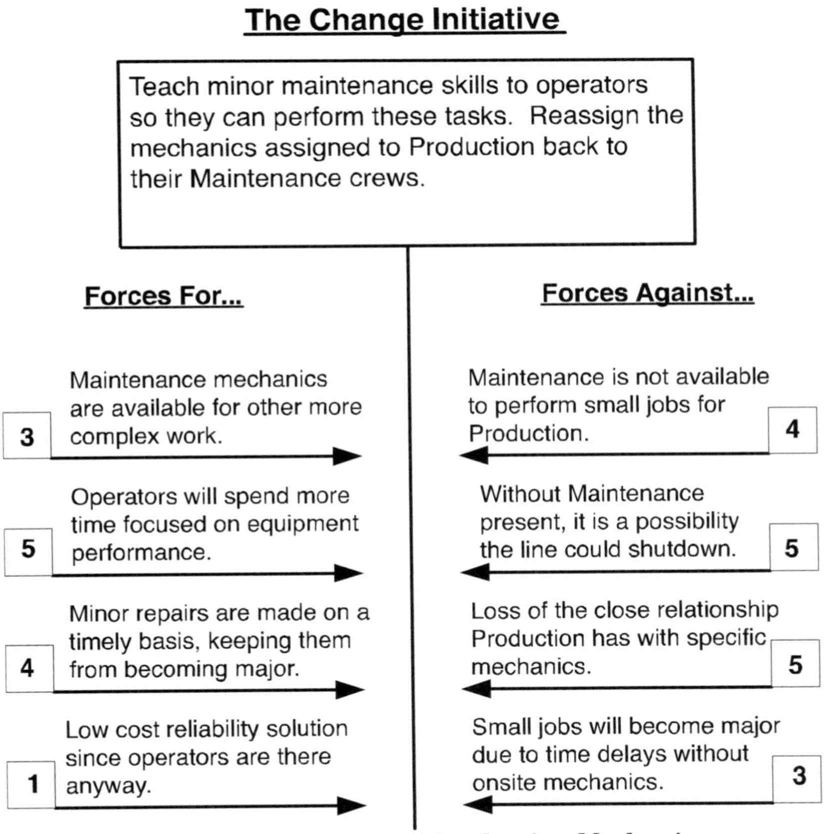

The Change Initiative

Teach minor maintenance skills to operators so they can perform these tasks. Reassign the mechanics assigned to Production back to their Maintenance crews.

Forces For...

Forces Against...

3 — Maintenance mechanics are available for other more complex work.	Maintenance is not available to perform small jobs for Production. — 4
5 — Operators will spend more time focused on equipment performance.	Without Maintenance present, it is a possibility the line could shutdown. — 5
4 — Minor repairs are made on a timely basis, keeping them from becoming major.	Loss of the close relationship Production has with specific mechanics. — 5
1 — Low cost reliability solution since operators are there anyway.	Small jobs will become major due to time delays without onsite mechanics. — 3

Figure 9-4 Elimination of Production Mechanics—
Force Field Analysis

Next the "forces against" were examined in detail. As you will remember, resistance is an emotional reaction to a change in one's status quo. By examining the "forces against" in detail, those present were able to see that the issues that they had with the change had been addressed in the work process design.

For example, the fact that maintenance mechanics would not be available for minor repairs was offset by the fact that the operators were to be taught the very same minor maintenance tasks. This was an even better solution than having the mechanics assigned to the production line; the operators were available 24 hours per day whereas the mechanics were only available during the day shift. When this was completely explained, the priority of 4 assigned to this force was reduced to a 1. This process was continued for each of the "forces against" to the point that Production changed their opinion of the initiative. In following this process, not only were the "forces against" reduced, but also everyone felt more comfortable with the initiative strengthening the "forces for change."

9.7 Resistance is Actually Group Learning

Resistance is not something to be overcome. Resistance is something to be addressed as part of the organization's learning process. As an internal consultant, your job is to facilitate this process so that the organization emerges from the effort stronger than when they entered. You need to recognize that resistance will take place. It has many forms that need to be identified. There are processes to address it, but once addressed as a learning experience, it can be addressed as part of the initiative.

One last thing. Because resistance is a naturally occurring part of change, don't take organizational resistance personally. It is not personal. Working harder to make your case does not solve the problem. Addressing resistance does!

Five Things to Think About or Do

1. Can you identify the various types of resistance as they appear on your change landscape? Re-read the section describing various types of resistance and see if you can find real life examples in your plant.

2. When you encounter resistance, what do you do about it now? Can you explain in light of this chapter how some of the things that are currently done at your plant to overcome resistance may not be correct.

3. Think about an initiative in which you are engaged that may be having problems. Draw a force field diagram. How you could enhance the value of the "forces for change" and diminish the affect of the "forces against."

4. Develop a plan to introduce the Force Field Analysis you developed in item #3 to the group with whom you are working.

5. Write out a short explanation and share it with others, describing how resistance is actually organizational learning and a valuable part of a change initiative.

THE INTERNAL CONSULTANT'S ROLE

To contribute value
With recognition for others

10.1 The ion Suffix

As an internal consultant, you will need to fill many roles as you practice consulting skills within your organization. The skills you will apply are similar in some ways to that of the external consultant. Yet your role is far broader and you bring to the effort skills that an external consultant will never be asked nor be able to apply. The reason for this is that you are not simply acting as an internal consultant for the organization. In many ways, you are also part of the organization for which you are consulting.

What does the suffix ion mean to you? Webster's dictionary defines it in two ways: "an action or process" or "a result of an action or process." In either case, the internal consultant's role in the process of change is to initiate, support, or follow-up with the organization as a result of actions or processes in which they are involved. For the internal consultant, ion is a very important suffix because the action words associated with it describe what is expected in this role.

Figure 10-1 The ion Tasks

10.2 The ion Tasks

1. Recognition

As you practice internal consulting for your organization, it will become clear to many that 1) you can deliver value to change initiatives that they wish to implement through your internal consultant skills, and 2) the organization is so involved in the day-to-day work efforts that they need you and the skills you possess to help them. This recognition may be the result of the work that you have done. Or it may result from recommendations from other clients as new change initiatives and how to accomplish them are discussed within senior management. In either case, recognition of your capabilities is a result of the quality work you have delivered in the past. You never know how this recognition will take place or where your next client will come from. For this reason, you need to carefully follow the work process steps outlined in Chapter 8, especially the completion and audit stages.

For example, I once worked on a project designed to enable the workers on the production line to gather equipment information electronically on hand-held devices. Prior to this initiative, they had written the information down on paper forms which were filed, but never reviewed. This initiative was largely driven by Maintenance. Because the information was gathered by the production workers, however, I got to spend a good deal of time with the senior mangers in Production. Later, after the information gathering initiative was successfully completed, I was asked to facilitate another production initiative completely outside of my area of expertise. I was asked based on the work I had done on the prior effort. You never know where your next client will come from, but positive recognition of your work will certainly keep them coming.

2. Preparation

There is nothing worse for internal consultants than to go to a meeting, a workshop, or any gathering of clients that they are running and to be unprepared. Every time you meet with an individual or group, your credibility is on the line. People expect that you will be prepared, maybe not to run the meeting (unless that is the role you are playing), but at least to provide the group with meeting structure and guidance related to helping them achieve the meeting's purpose.

Preparation also applies to one-on-one meetings with your client. Although they do not expect you to initially understand their problems or longer-term vision, you are expected to be prepared for these meetings. Because you are internal to the organization, you have an advantage. You can do some research before your meeting so that, at the minimum, you understand the issue and maybe even have some preliminary ideas about how to proceed. Other types of preparation for these meetings include plans for conducting the work initiative if you are given the assignment. Chapter 8 explains the process in detail. It should be applied to every initiative in which you get involved. Therefore, coming to the meeting with your client with this framework in mind will show them that you have adequately prepared.

Preparation also applies to every phase of the work initiative in which you are involved. When people come to a meeting, work in groups, or even make presentations regarding the next steps of the change initiative, your preparation or assistance in getting others prepared makes them feel confident about the work they have done with you as their internal consultant.

3. Coordination

Almost all change initiatives involve groups of people working together to implement a new process, strategy, work direction, or even a simple way to conduct business in a different manner. As an internal consultant, you can play an important role coordinating the activities associated with the change initiative.

Although coordination may seem like a menial task for the internal consultant, it is not! Coordination of a change initiative has internal consultant implications beyond simply handling logistics and tasks that enable the effort to move forward. Coordination serves to position you as a central and key part of the process. This doesn't mean you have to do all of the work yourself—the work can be delegated—but it does allow you to be seen in a central role. A manager I once worked with told me that to serve the group is a powerful way to make your presence known. He was right; coordination is one way to accomplish this effort.

It does not only allow you to be seen by those involved as a key player. It also helps establish your larger and more important role in meetings where you can provide guidance and help the organization reach their stated goals.

Another important part of coordination is that it allows you as the internal consultant to keep the process moving by making sure that things happen and people accomplish their action items. All too often, people make commitments at meetings but get sidetracked during their daily work. As a result, the things they committed to do not get done. In a coordination role, you can reduce if not eliminate these problems. The other team members will appreciate this because it helps keep the effort on track for all those involved.

Tasks that fit within the scope of internal consultant coordination include:

- Preparing the meeting agenda and sending it out before the meeting

- Setting up the meeting including the location, audio-visuals, and other things required
- Making sure everyone attends
- Making certain that meeting notes with action items are completed in a timely manner
- Following up on action items to make certain they are completed on schedule

Although you may feel that the team should take ownership and be responsible for these tasks—after all, you are only there for support—this does not always work. A good coordination effort builds internal consultant credibility and pays off in many other areas of the internal consultant effort.

4. Facilitation

This is another role of the internal consultant. There is a vast difference between facilitating and running a meeting or working session. When facilitating, you spend the majority of your time monitoring the meeting's work process. For example:

- Is everyone participating?
- Are people listening to what is being said?
- Do people have sufficient time to express their opinions?
- Are there multiple conversations going on which distract the team from the task at hand?
- Is time being used wisely by sticking to the agenda?
- Is someone monopolizing the conversation?
- Is the agenda being followed and, if not, how do you get the team back on track?

None of these tasks have anything to do with actually running the meeting. A good internal consultant will allow the team members to run the meeting; they will then focus their attention on work process so that they can help the meeting be effective.

Just as good facilitators allow the meeting to be run by the team; bad facilitators do just the opposite. They take over and run the meeting themselves to the detriment of the team. Being in charge of the work of the team is not the role of the internal consultant. This behavior serves to lessen the value of the meeting and the team's effort; it also slows down the development of the members in acquiring sound work process skills.

I was once involved in setting up a process where the plant was divided into business teams by production line. Each team had representatives from all departments which were involved with the operation of that specific line. For a brief time, Production ran the meeting (which met once per week) in order to focus the team members away from their day-to-day activities toward work on strategic improvements for their area.

These meetings worked well until management decided that team should be facilitated by members of the plant manager's staff. With this addition to the team structure, many of the teams still functioned well—including those areas where the managers did not know the functioning of the line and, as a result, were forced to work as a true facilitator. Where the managers did understand the line, they were quick to add their ideas to the discussion; by virtue of their position, they took over the meetings.

5.   **Education**

Another role for the internal consultant is to educate those for whom they are providing consulting services. This education can be provided in different forms: 1) directly through classroom training or workshops or 2) indirectly through group or individual coaching. In either case, as an internal consultant, you have the skills and information you need to impart to your clients. Education is essential; in your role as internal consultant, you need to leave your clients better prepared for sustaining the current initiative and to be able to handle the next initiative in which you may or may not be involved.

Direct training via training sessions or workshops has the potential of being effective when the entire group needs to collectively learn a skill. You need to be sure that this type of training is really required because it interrupts the process in which the group is engaged. It also can be expensive. If it isn't exactly the right training, it can also hamper the group's development.

Suppose that you are consulting with a group attempting to develop goals, initiatives, and activities related to the vision established by the maintenance manager. The process that the group should be using is called the Goal Achievement Model (see my book *Successfully Managing Change in Organizations: A Users Guide* or Appendix 1). This model is designed to help a group perform this activity. However, because most people have never used it, it needs to be taught to the group and collectively understood; hence, the need to interrupt the work process and teach the concept.

There are other times when formal training by internal consultants is not needed. In these instances, internal consultants can impart the knowledge through training of the individuals so that they understand the concept that the internal consultants want them to employ. This assumes that other group members understand the concept and are not in need of this training. This work should be undertaken outside of the group setting. Otherwise, those receiving this training could feel as if they have been singled out. Individual training in a group setting could be embarrassing. If so, the point that the internal consultant is making will be lost. It also will cause those who understand the material to become bored and lose interest, which is often hard to regain.

There are times when individual training within the group setting is needed and can't wait until a later time. This occurs when it is obvious to the internal consultants that the individual clearly does not understand a point and is disrupting the progress of the work group. Even in this situation, when immediate training is needed, the internal consultants should call for a break and handle the individual training in private.

6. Documentation

Every group has a need to document their meetings, activities, and action items. Even while assuring that this documentation takes place, doing the actual work should not be a primary role of internal consultants. It has been said that you can not facilitate and take notes for a meeting or work session at the same time. I can attest to this statement, having tried and failed at handling this dual role on numerous occasions. As the internal consultant, you need to make sure that each session has a scribe and that the meeting notes and, even more important, the action items from the session are correctly captured. These notes are important because they document what has been discussed as well as the work that individuals have to do between meetings.

Some groups have a rotating scribe, a different person handles the task on a rotating basis. This can work but it is not as good as one person who is able to handle this task correctly, always taking the notes. A single scribe will be able to maintain consistency; those who like to do this type of work will feel as if they are contributing. Make certain, however, that the scribe also participates. Otherwise, potentially valuable input may be lost as the person is relegated to the note-taking task.

Even though I said that as an internal consultant you should not take notes, you should jot down key items and action items. Then during the meeting recap, you can make certain that these are captured in the documentation.

You should also list items that are commonly referred to as "parking lot" items. These are discussion points that do not specifically relate to the immediate discussion, but they are points that you do not wish to lose. Maintain a list of these items on a separate sheet of paper or flip chart so that you and the group can discuss them at a later date.

7. Notification

Another critical role of the internal consultant is notification. Although you may think this applies to notification about meetings, that is not the case. (That communication is covered in coordination.) Notification deals with the action items that result from group or individual meetings and discussions. These are the critical elements that keep an initiative alive and moving forward.

All too often, action items are accepted during working sessions, but get quickly forgotten. They are then only remembered just before the next meeting takes place, at which time the person with the assigned action items rushes to complete the assigned tasks. The result of this last minute effort is 1) hastily completed work which is often less than the individual doing no work at all, or 2) work which doesn't get completed on time.

The latter alternative not only delays the progress of the work, but also gets the other group members who have completed their tasks angry at those persons who did not do their assigned work. This has the potential of causing a project breakdown. It needs to be avoided at all costs. Thus, internal consultants have a notification role. It is their job, between working sessions, to periodically contact those with assigned action items to make sure that the work is being done in a timely manner. This is a simple task, but one which is highly important and often overlooked.

8. Cooperation

Working together as a team takes a great deal of effort. This effort is even more pronounced when the group is working on a change management initiative. It is, therefore, the job of the internal consultants to assure that those in the group cooperate with one another for the collective good of the team and the work initiative. Cooperation is a sub-task of facilitation, but is listed separately because of its importance to the process. When people cooperate with one another, the overall effort is strengthened; the outcome is far better than any one of the members could have achieved individually. On the other hand, when there is a lack of cooperation, the success of the effort is lessened. Lack of cooperation could lead to outright failure, even though the individual members want it to succeed.

Lack of cooperation fits the resistance model previously discussed. Usually people who are not cooperating are either 1) holding back and not participating or 2) taking an opposing position, challenging everything that is decided.

Holding Back (Lack of Participation).

This behavior can run the whole gamut from not showing up at team meetings to showing up, but not participating. As the internal consultant, you need to understand why the individuals are behaving in this manner and address their issues. If they are missing meetings, they need to understand the need that the team has for their presence and recommit themselves to becoming part of the group. If they don't want to attend and don't want to be part of the process; they need to be replaced.

At the other end of this spectrum are those who want to attend and do so, but do not participate. There are many possible reasons for holding back in this manner. They may not want to be part of the process, but feel they would be punished if they spoke out. They may feel threatened and not adequate for the task. In either of these cases, or others, it is your job as the internal consultant to talk with these people and resolve the situation. Often a simple discussion will change a non-participant to someone actively involved. In other circumstances, you may conclude that the best interests of the effort require their replacement.

Taking the Opposing Position.

People who continuously oppose the positions of a work team are often referred to as acting as the "devil's advocate." At times you need someone who is willing to take an opposing position. It forces the team to think about the justification for their ideas. But people sometimes take this to an extreme; often because they like being the center of attention or simply because they really are openly resisting the effort. As the internal consultant, you need to quickly take corrective action when this behavior is observed. It undermines the team's efforts and breaks down the cooperation between members.

9. Frustration

It is a major task for internal consultants to help minimize frustration of individuals and of the work group. As projects and work initiatives progress, there are many who will become dissatisfied or frustrated with the progress being made or even with the other members of the team. They would simply like to get things moving a lot quicker than the apparent progress which they observe. Often they have a "quick fix" solution to the lack of progress and try to push their approach onto the rest of the team members. The problem with this is that if you carefully examine what they propose, you will quickly discover that it is not really a "quick fix" solution, but one that would severely cripple the work already done by the team.

In the reliability / maintenance business, things are very complex; it is virtually impossible to mitigate frustration with a "quick fix" solution. Suppose that you and your team are working on an initiative to develop and implement an improved planning and scheduling process. Of course it is progressing slowly because you need to build and get buy-in for the new process, train those involved, run a pilot effort to validate the process, deploy it across the site, and put in place mechanisms to ensure sustainability of the work process you are implementing. Then one of the managers suggests simply deploying the process without the pre-requisites of training and the other essential ingredients of success being in place.

This idea of "just deploy it" is evidence of their frustration with the slow pace to date. As the internal consultant, you clearly recognize that employing a "quick fix" solution to resolve the manager's frustration will cripple the effort and possibly cause premature failure. This is where you need to handle the frustration and explain the harm that a "quick fix" resolution will cause.

When discussed in light of the potential damage the "quick fix" will cause, most people will relent, allowing their frustration to dissipate and the effort to proceed at its correct pace. For those who do not recognize the problem they are about to cause, you may need to bring in your client or someone even at a higher level to bring the situation under control. This should be done as a last resort. But in order to fulfill your responsibility to the work and your ultimate clients—those who will be receiving the end result of the work—you can not allow those who advocate the "quick fix" to derail the process.

The old adage taught to me by a close friend is that, in the world of change management, "slow is fast." The problem is that people used to working in the hectic world of daily reliability and maintenance don't realize this fact. Why should they? Most work in reactive work settings where fast is not usually fast enough. Your job as an internal consultant is to get them to slow down and do it right the first time.

10. Motivation

This aspect of the internal consultant's role has two phases. First you need to be able to keep the work group motivated as they work through the process of bringing change to their plant or company. Second, you need to keep yourself motivated as you progress along the difficult path which you have chosen as your career.

Motivating Others

The members of the work group require individual motivation. Their work experience is most often totally based on reactive maintenance and addressing the problem of the day. They are so focused on this type of work that they often have little time to think less about work in a strategic fashion. Working strategically is different. It requires engaging a different side of the brain and most certainly involves working in a very different manner.

For an individual, this type of change is even more difficult if you are going to move from your regular job to working full time on a strategic change initiative.

At one point in my career, I was assigned to a full-time work team that was developing a reliability project. There was such a large volume of work that it required those on the team to work full-time. In addition, the work was centered at corporate head-quarters; I moved from a plant work environment to corporate with a different office location, working hours, dress code, and people whom I had never met. This transition was difficult for me. Up to this point, I had been focused on fixing plant problems as quickly as possible. As the work effort progressed, I could see things getting done, but it seemed to be taking an inordinate amount of time to make progress.

It was at this point that I felt I had lost my motivation to do longer-term strategic work. I desperately wanted to return to my former line job. The consultants who were working on the project were able to show the team how an effort of this sort had added immense value to other companies with whom they had worked. They were also able to show me how the work that our team was doing was making progress towards our goal. Without any prior experience with this type of work, I had not been able to see the goal clearly nor the progress. The consultants motivated me to continue and re-energized my efforts.

Motivation is even more difficult when the work team is not full-time and they have to balance their daily job with that of the change initiative. In this situation, they are constantly being pulled in two directions. First they need to address their daily work, which often includes keeping the plant running. Second, they need to be able to switch hats and work on a long-term strategic initiative with its vastly different requirements. As the internal consultant, you need to carefully monitor the level of motivation of the team as a whole and of the individuals. You can sense when the group or an individual needs to be re-motivated simply by watching how the work is progressing. Some symptoms include:

- Missing meetings due to the demands of their regular job
- Showing up late for meetings

- Taking cell phone calls during working sessions
- Questioning the value of the work
- Not completing action items
- Not focusing on the agenda or task at hand
- Expressing negative feelings about the work
- Explaining why "management will never let us do this"

Motivating Yourself

The second part of motivation is self-motivation. As the internal consultant on the job, you will not only have to motivate the team as the consultants did for me, but also you will need to motivate yourself. This is easy in the good times and difficult in the hard times. Work initiatives see both. In many ways, you are the role model for the team as they work their way through the process. You have seen change efforts succeed as well as fail; you need to use your experience to help the group see the big picture and value-added result that they can achieve. They only see the work that they are doing. You see the bigger picture and it is your task to help them see it as well.

11. Appreciation

All too often this part of the internal consultant and management's job goes unaddressed. Both you as the internal consultant and management—specifically the sponsor of the initiative—have a very important role in this area. In fact you have two roles to perform. Management's role is to provide appreciation and recognition for a job well done; you have the same role as an internal consultant. But your level of appreciation also addresses showing appreciation for people being open to new ways of thinking and working. This other role is to assure that, if management is failing to show their appreciation, the people you work with are still encouraged to change their ways.

Appreciation is a powerful tool, but it must be used in a balanced fashion. If there is too little appreciation, then those working on the change initiative are not 100% certain that what they are doing is correct; they have no assurance that their work won't simply be rejected off-hand when they finish. Too little appreciation for the effort is also a de-motivator for the group. Appreciation with reinforcement of direction shows that the work is on track and the sponsor is encouraging the group to continue.

Too much appreciation is just as bad as not enough. I once worked for a manager who told me that everything I did was "excellent." Even I knew better than that. This level of appreciation was bad because I didn't know where I stood any better than if he had said nothing. As an internal consultant, you need to make sure first that appreciation takes place and second that it is balanced so that it has meaning and value.

12. Creation

All things considered, the creation of new direction, new initiatives, and new and improved ways of working is what the role of the internal consultant is all about. For those of us in this role, there is nothing more rewarding than to have successfully navigated all of the ion suffix action tasks and deploy something that, in your heart, you and the work team know will improve the business. Also as an internal consultant, you need to recognize that not only have you helped to create something new and innovative for the business, but also you have created new knowledge and understanding in those who have made the journey with you. This unto itself can be even more rewarding than the new work process; those with this new knowledge can now replicate it in many other ways. These new behaviors breed new learning in others. Ultimately, the entire organization improves.

13. Continuation

Sustainability of a change initiative is absolutely important to an organization. That is why, in addition to listing continuation as one of the ion tasks, I have dedicated all of Chapter 15 to this very important topic.

There is probably nothing worse or more depressing than to spend time and energy on a change initiative, and then find out six months after deployment that it has failed. Sometimes if you wait this long to check on the status, the organization may even have forgotten the initiative ever existed.

People who work on projects consider them as linear events—efforts with a specific start and completion date. That is not the case with change initiatives. They need ongoing care and feeding especially in their infancy, if they are going to survive over the long term. Therefore, a critical part of any initiative needs to ask "how are you (the organization) going to sustain what has been developed into the future?" As an internal consultant, failure to address this issue on behalf of the team is a failure to address the task of continuation.

I was once involved in a reliability effort designed to improve how the maintenance organization interacted with their production counterparts. For a number of reasons, the effort was poorly designed and poorly deployed. Even so, everyone tried to make it a success because they respected the manager who was the initiative's sponsor. Based on this, it survived for several years. At the end of that time, the manger retired. When he left, the initiative's driving force left as well and the process ended.

It was a shame. The effort had value, but no thought was put into how to sustain it after the sponsor left. Imagine how he would have felt had he visited the plant six months after his retirement and discovered that the effort that he had put so must time and effort into no longer existed. Efforts in the area of continuation can help prevent this type of unfortunate outcome. It is your role and that of the work team to address this issue so that the process of change continues.

As you can see, the tasks associated with the ion suffix are numerous. They represent a set of internal consultant tools that add immense value when employed properly. Conversely, you can also see how missing one or more of these tasks can cause a change initiative to fail to achieve its potential or simply fail completely.

Five Things to Think About or Do

1. List all of the ion suffixes and then write why they are important to you and how you plan to address them.
2. Select an initiative that succeeded and identify the ion suffixes that you executed well. If you addressed all of them, you are doing well. If you didn't, consider those that were missed. What could you do next time to make sure that they are included?
3. Select an initiative that failed. Identify the ion suffixes that were missed. If you and the change team had addressed them, would the outcome of the initiative been different?
4. Develop a mini training course using PowerPoint so that you can explain the ion suffixes to your client and work team.
5. Teach your mini training course to a client or work group so that they understand and can support the application of the ion suffixes as part of the change process.

WORK TEAMS

The value of the input from the many
Far exceeds the value of input from the individual

11.1 You Are Not The Problem Solver

As an internal consultant, you can not simply show up at a work site with your answers to the site's problems, deliver them, and expect that they will be implemented. First, without detailed knowledge of the issues and the resultant problems, you may miss the mark. But even more important than that is that, as an internal consultant, you are not there to solve the site's problems. Rather you are the person who will help them solve the problem for themselves.

Your job is to support the gathering of the data. Then, working with the site personnel, you help clearly define the problem and develop a viable solution. You could accomplish this task by talking individually to each person on site, summarizing the information, developing a solution, and then reviewing it with all of the site personnel to reach consensus. Once this set of tasks was completed, you could then work with the site personnel to deploy the solution. This is a very time-consuming task, one which, if attempted, would probably never reach completion.

Another solution is to gather knowledgeable representatives from across all of the affected organizations, then solve the problem as a team, with you as the internal consultant providing guidance and support. This is a much more practical and efficient way to solve a problem. In order to be able to successfully accomplish this task, you need to clearly understand teams and how you as an internal consultant fit.

11.2 The Definition of Teams

This chapter is about teams and your role as an internal consultant with these teams. The focus will be how to make teams work—either where they have never existed before or where they currently exist—but not in the way that creates or adds value to the organization. First we must establish exactly what a team *is* and why teams should be used in everyday business. In *The Wisdom of Teams* (New York: Harper, 1994), Jon R. Katzenbach and Douglas K. Smith define a team as follows:

> "A team is a small number of people with complementary skills who are committed to a common purpose, performance goals, and approach for which they hold themselves mutually accountable."

I have underlined five key aspects of this definition, all equally important, to discuss separately.

A Small Number of Ppeople

Some people think that the size of a team does not matter. They think that if you want people with a common purpose to work together, then you simply need to get all interested parties together for maximum benefit. Yet size is important. If a team is too small—less than four—it probably will not be of sufficient size to represent all of those who would be affected by its work. Furthermore, such a small group most likely will not have enough membership to accomplish the task. On the other hand, if a group is too large—greater than twelve—it could be too large to get anything accomplished. My own opinion is that the optimal size for a team ranges from four to twelve. Size is important to consider when you are creating a team because you want positive results, not frustration and failure.

Sometimes, getting the right number of people on the team is very easy. However, you may run into the problem of not having enough or having too many members. What can you do? If you don't have enough members, consider getting two or three representatives from each affected area in order to increase the team's size. If this solution isn't workable, then consider getting people from other groups that may not be as directly affected, but that still have a stake in the issue. Use them to increase the numbers. When you are short the number of members you need, then check that you have included all groups that are affected. Often the issue of size will resolve itself when you recognize that you have inadvertently left out a group that should have been included.

Shortage of team members is usually not the problem. The problem usually is the opposite: having too many people for the team. Often when you are working on an initiative that will affect your company, many groups need and want to be included. The result can be that you have a team of more than twelve. You can apply two possible solutions to this case. The first is to divide the problem into parts, then create more than one team. The benefit is that you have groups of manageable size. The risk is the possibility that the smaller groups arrive at conflicting solutions. Bringing the groups back together on a periodic basis to validate what the others are doing can mitigate this problem. You can also have group representatives meet and agree on a common direction, thereby minimizing this risk.

The second solution is to start with one large team, then break it into smaller subgroups to work on parts of the problem. Periodically the small groups are brought back into the large team format. Consistency of approach can then be maintained.

In the first of these approaches, you start with many small groups and bring them together to work out issues. In the second, you start with one large team, and then break into smaller groups to do the work. The first approach can be used when consistency between the groups is not a major concern, the second approach when consistency is important.

Complementary Skills

Another important aspect of a team is that the members have complementary skills. The collective skills of the members must provide what you need to accomplish the team's purpose. The old adage that the whole is greater than the sum of the parts is most certainly true in successful teams. Every sport played by a team demonstrates the power of complementary skills; individual skills that when taken together create a team that can accomplish the desired outcome.

Committed to a Common Purpose

If a group of people are trying to accomplish something that they collectively want to achieve, then they are committed to a common purpose. The area where a problem most often occurs is in the definition of "commitment." How do you define or know the level of commitment of individual team members? How do you determine the level of commitment that is necessary? These questions are not easily answered because commitment comes in varying degrees. It is determined by the team and, in most cases, defined by the team's specific assignment at any given time. A good team is self-regulating. If it is a mature team, its members will identify the level of commitment that is required.

Suppose a team is working on a critical business problem. The team must meet four hours per day for a week. People who are late for the meetings or don't provide the necessary level of input are likely to have their behavior addressed by the other team members. The group needs must be fulfilled. At times, a request for additional commitment may be made in a joking way. Sometimes an outright demand for more commitment may be necessary. In the end, the team will try to regulate from within the needed commitment from its members.

Performance Goals and Approach

A team can not be a team unless it has something to do. Business teams usually have goals or initiatives and activities on which the team can focus its attention. Although the team may start at the goal level, it quickly moves through the stages of the Goal Achievement Model and actually works at the initiative or activity level. It is here where teams can create value for your company.

The specific approach that a team takes will vary depending on the assignment. In all cases, they must follow a common approach within the team so that their time is optimized. Attending a meeting with no agenda and no process, or one where the meeting doesn't flow, frustrates team members and wastes their time. A good team does not allow this to happen; it is focused on results.

Mutually Accountable

Mutual accountability is another factor important to a team's success. Certainly teams and their members are accountable to a higher level within the organization. Equally important is the fact that the members are accountable to each other. Each member brings a separate complementary skill to the team that requires the members to depend on each other.

This aspect of teams also points to a potential problem. Teams are often built across functional organizations; the members are from different departments and work groups, and even from different levels of hierarchy. As the individual members go through the stages of becoming a team, the members bond; they become mutually accountable to one another. When the team's assignment is a project with a finite end point, then the problem of accountability can be minimized. By the time the members get to conflicts between their own regular responsibilities and the team's, the project is usually coming to an end.

What about the long-term efforts? Teams that operate continuously effectively create a matrix organization within the company. Thus you would be part of a regular work group and have an immediate superior. At the same time, as a member of a team, you may be supervised by someone outside of your immediate organization. In this sense you have two bosses. This is not a problem when the goals and initiatives of your immediate supervisor and the team are in alignment. But what if they are not in alignment? This conflict can create a problem for your personal performance, your work group, and the team.

When you are pulled in two different directions, which way do you go? Unfortunately for teams in this situation, the members usually align themselves with their immediate supervisor, the one who pays their salary. In turn, problems develop for the team, inhibiting its ability to do its work. This conflict speaks to the need for alignment. Then what you do for the team will also complement what you do for your primary job and work environment. Without alignment something will suffer—and it usually is the team.

11.3 Types of Teams

Teams typically come together to solve problems. These problems fall into three categories; short term, long term, and on-going. The first two are characterized by defined end points; that point at which the problem is to be solved and the solution deployed. The other type of problem is one that is on-going such as a change initiative. These efforts have solutions but just as soon as you solve and deploy one aspect, others emerge so that the scope of the overall initiative and the resultant benefits continue to grow.

Short-Term Efforts

The duration of the allotted time for the team to do its work dictates the type of team and how quickly the team has to reach its full level of effectiveness. It also dictates how you as the internal consultant need to work with the team. Figure 11-1 describes an initiative that has a short duration before its expected deployment. As a result, the team needs to attain its maximum effectiveness

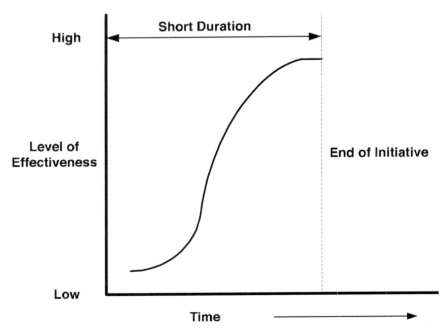

Figure 11-1 Short Duration Effort

over a very short time horizon. This requirement dictates the type of team that must be employed. In this case, the team needs to have full-time members reassigned to the team for the duration of the effort. For some organizations, this has a significant impact on their ability to accomplish the work required to operate the plant. Having key members of the organization unavailable when resources are scarce can cause real problems. Organizations deal with this in two ways. First they assign the resources to the team and make the commitment to suffer their absence until the work has been completed.

Long-Term Efforts

The second alternative is to recognize that the duration assigned for the team to conduct the work is too short and extend it. This sets in place a condition that allows key resources to work on the initiative part-time with the balance applied to their regular job. This approach, shown in Figure 11-2, will take longer, but

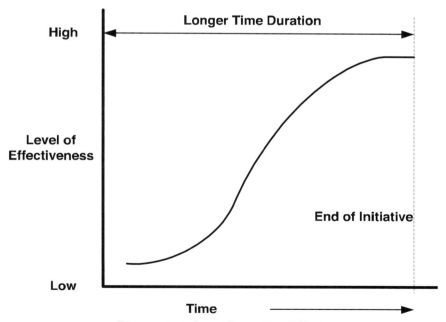

Figure 11-2 Long Duration Efforts

allows the team to accomplish its goal and to reach its full level of effectiveness over a longer period without negative consequences to the day-to-day work.

On-Going Efforts

The third type of team, as shown in Figure 11-3, is one where the team is focused on organizational change. Initiatives of this nature have no time restriction for the overall effort because change is an ongoing process. They may have restrictions on the sub-initiatives, but even these are somewhat flexible.

Effectiveness for teams working on change initiatives is cumulative, building from sub-initiative to sub-initiative. As you can see in the figure, team effectiveness related to a sub-initiative builds slowly, picks up speed, and then levels off as the sub-initiative is completed. The difference in this diagram is that when one sub-initiative ends the team has learned new things and a new sub-initiative usually begins.

Suppose a team is developing an improved planning process. As they reach the end of the development phase of the process,

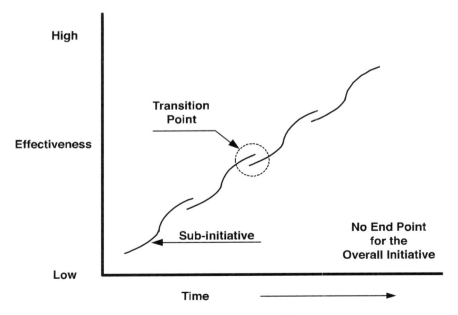

Figure 11-3 On-Going Efforts

they recognize that in order to have a successful process they need to provide adequate planner training. This launches the team on a new sub-initiative to develop and deploy planner training. As this efforts nears completion, the team recognizes other sub initiatives that need to be addressed. In this manner, the team learns about the complexity of the overall task; their level of effectiveness continues to build over time and multiple initiatives.

In this scenario, one would think that the amount of time team members would spend working on the initiative could be drawn out with even less frequent meetings than described in Figure 11-2. Not so! Obviously the team does not have to be assembled and work full time. But you can not allow extended time between meetings or the proper level of effectiveness will never be obtained. Think about what happens at your own site when you have a task and meet infrequently. Usually everyone forgets about the action items they promised until just before the meeting. Then everyone is scrambling around trying to fulfill their commitments. The result is that the team members do not deliver the quality for which they are capable.

As the internal consultant in all three scenarios, you have a great deal of responsibility. For short duration initiatives, you need to work closely with the team and help them attain maximum effectiveness over a very short period of time. For longer-term initiatives, you need to help them attain effectiveness. You also need to support team continuity because they are not working full time on the effort. Your role is even more critical for the change related initiatives—those without an end point. In this case, you not only need to build effectiveness and assure continuity, but you also need to assure that these vital components of team performance are maintained over the long term as members leave team and new members (without the team history) arrive.

11.4 Membership

One role that you have as an internal consultant is that of establishing effective membership on whatever team you are trying to build. When site leadership recognizes that the initiative requires a team focus, they will want to build the team quickly and get on with the work at hand.

This rapid creation of a work team is not always the best approach. Teams need to be created with some forethought; this is where you play a significant role. Left to their own devices, senior management will often delegate setting up the team to their subordinates. They usually do this because they have many other pressing tasks at hand. In addition, although they know the overall capabilities of those in their organization, they don't know the specific capabilities that are needed by the team nor who actually is best suited for the assignment.

This approach has the potential for being an adequate solution. However, the directive to "get on with the work of the team as quickly as possible" can cause those making the selection to pick the wrong people. In addition, the subordinates need to be as committed to the effort as their leadership. If so, then they will appoint people who they believe will add value to the effort in spite of the fact that these very appointments may leave them short-handed for an extended period of time. If they are not committed, or possibly never told the rationale for the team, they will try to mitigate the loss of resources. The team may then have people assigned who are not appropriate.

As the internal consultant involved with this effort, you should work with those who are selecting the team members to make certain that the best people are assigned to the task. Different approaches can be taken based on the type of effort, e.g., short duration—full time, long duration—part time but with frequent meetings, and a change-based team which will meet frequently and over an extended time frame.

When working with those making the selection of team members, consider the following:

- Attempt to have the senior management make the assignments and not delegate the task to subordinates who are faced with a difficult choice: either support the effort with the best people and short-change the day-to-day work or keep the best people and short-change the initiative.

- Select people with complementary skills who understand and support the effort. Having people with skills that support one another is fine, but having those skills possessed by people who can see the value of the effort is far better.

• Select a mixture of people with transitional and transactional leadership qualities. Those who are transitional will be able to help develop the long-term vision and approach. Those who are transactional will be able to help develop the way to implement the effort over the short term.

• Select people who have high levels of credibility with those in the organization. This will make the work that is being done by the team easier to deploy; people feel comfortable adopting new ways of doing things if they have been developed by people who have their respect.

• Make sure that the organizational impact imposed by those selected is understood. Mitigation strategies should be in place so that the work of the team doesn't suffer nor does the work at the plant.

All of these requirements for selecting team members can be influenced by you in your internal consulting role. It is important that you apply this influence because the people selected are those with whom you will be working.

11.5 The Internal Consultant's Role in the Process

Once the team is formed, your job as internal consultant really begins. You need to bring the team together and help them identify why they have been assembled, what they are being asked to accomplish, and what is the time frame in which they need to complete the work. Changing people from a group of individuals meeting to complete an assigned task to a team with "common goals and mutually accountability" is no small achievement. They may know why they were assigned, but certainly the detailed knowledge about what it is that they are to do is not well understood. Each person brings their own interpretation of the end result to the group. As the internal consultant, you need to help channel this towards a common purpose.

Therefore, your first task is to have the group arrive at a clear agreed-upon vision of what the end state of their effort will deliver to the company. This vision needs to be stated in such a manner that each team member must hold the identical, crystal clear picture of this yet-to-be-achieved future end state. This consensus may take some work on your part; this discussion needs to be carefully facilitated. Because all members have their own view, it is easy for this discussion to get off track or immediately jump to trying to determine detailed solutions.

With the vision set, you next step is to work your way through the Goal Achievement Model. This model enables the group to establish a vision, then work their way through goals, initiatives, and specific activities to accomplish the work at hand. With the Goal Achievement Model developed, each person working on their specific assigned activities can clearly see how these activities support the initiatives, how the initiatives support the goals, and ultimately how the goals drive the vision.

The Goal Achievement Model is described in the appendix, which has been extracted for your use from my first book Successfully Managing Change in Organizations: A Users Guide.

Building a vision, goals, initiatives, and specific work activities may seem easy, but it is not. This process takes a considerable amount of time and effort to arrive at what the team feels is the correct answer for the plant. Along the way, team members learn a lot about themselves, the other members, and what it feels like to be a team with a common purpose. As the internal consultant facilitating the effort, you can:

- Focus your efforts on the work process that the team is employing to accomplish the task. You can then help the team stay on track and help promote ownership of the final product.

- Teach the team the tools that they need to perform their assigned task. These tools include the Goal Achievement Model, the eight elements of change, and the four elements of culture.

- Monitor team logistics, including agendas and meeting notes. Taking on these tasks helps the team focus on what they are developing without worrying about logistical details. It also provides you with some level of control over the effort.

- Control the use of team time—address lateness or non-attendance concerns. Both of these elements show lack of commitment to the effort. Because you are trying to build a team with a common purpose, perceived lack of commitment needs to be quickly addressed.

• Objectively address team process issues that are not working, either with the team or with individuals. Because the team members are totally engaged in the work effort, they may not even recognize that work process issues are not functioning as they should. It is your job not only to recognize them, but also to be responsible for timely corrective action.

• Ask questions that help the team to address issues in a comprehensive manner. Open-ended questions often enable the team to discuss issues without having them appear to be confrontational.

• Don't allow the team to be swayed by a dominant member. This is always something to watch for because 1) dominant members try to take charge and control the outcome, 2) less dominant members often let them and don't have the opportunity to add the value that they could, and 3) the end result is not what the team would have developed collectively. When team decisions are made by dominant people, others will not always commit to the implementation. This will set the effort up for failure before it even starts.

• Make certain everyone is involved. See that everyone participates.

There are also things that as an internal consultant you need to avoid when working with teams:

• Don't provide content to the team unless asked or you have something of significance to offer. (In either case, ask for permission to change roles. Then provide your input as a team member and go back to facilitation.)

• Don't try to run the team. That is not your role and will go a long way to undermining team ownership of the end product.

• Don't confront problems involving an individual's action in front of the team. (Provide corrective action in private and praise in public.)

11.6 Teams and the Eight Elements of Change

For teams to be successful, they need to understand and integrate the eight elements of change. Failure to work in this manner will leave gaps in whatever it is that the team develops. This will set them up for failure in their assigned tasks as well as in their deliverables. The following shows why the eight elements are required.

- Leadership. The team needs leadership to provide direction as they work through their assignment. Different members may lead at different points along the way, based on their individual levels of experience. The internal consultant should not be the leader.

- Work Process. Without a defined process, the team will never accomplish the assigned task. The work process is what helps the team work through the effort and arrive at the end with agreed-upon deliverables.

- Structure. Without structure the team will have no framework in which to act. This element also includes team roles, such as timekeeper and scribe, which may be rotated among members.

- Group Learning. Without group learning the team will not be able to learn from their efforts nor from the efforts of others with whom they will interact in the performance of their assigned task.

- Technology. Technology in this context deals with the systems applications that support the work and / or the deliverables of the team. In some cases, this may only be a small component of the team's work. However, in others—such as the replacement of a computer system—it may play a major part.

- Communications. This element is critical. Effective communication is very important to assure common understanding of the work required and the functioning of the team. Failure of the communication system will quickly lead to failure of the team.

- Interrelationships. This element is equally as important as that of communications. For teams to function well, they need to have positive interrelationships. Otherwise, the internal consultant will be very busy trying to hold the team together and nothing will be accomplished.

• Rewards. Every team needs reinforcement to show them that the work they are doing has value and is worth their time and effort. In this case, rewards seldom equate to money. Rather they are provided by recognition and support of the effort from senior management.

11.7 Teams and the Four Elements of Culture

The four elements of culture are also important to teams as they work their way through often difficult change-related assignments.

• Organizational Values. The work of the team must be aligned with the organization's values. If not, there will be a serious mismatch between what the team delivers and what the organization truly believes they are supposed to be doing as they execute work. Suppose the team is working to deliver a detailed preventive maintenance program, however the site's value system is totally reactive. In this case, there will be problems with successfully deploying the initiative due to a mismatch between the team's deliverables and the values of the site. Before a team begins working on an initiative, they need to be sure that it will be supported by the existing value system.

• Role Models. The role models in an organization are those people who the rest of us look to when we wish to identify what success looks like. By emulating these people, we hope to attain the same level of success. Not everyone on the change team will necessarily be a site role model. In fact, few of the existing role models will be on the team. The reason is that role models depict success in the existing culture whereas the change team's task is to alter this status quo. However, the people on the team are people of influence and have credibility in the organization. Therefore, it is incumbent upon the team to 1) identify the existing role models and get them aligned with the change initiative and 2) determine ways that team members can become role models for the new process.

• Rites and Rituals. As the work of the team is deployed,

the rituals that the site has employed in the past will undoubtedly be altered. This doesn't only impact how things are done, but also how behaviors are reinforced. Therefore, as the team deploys changes in the rituals, they need to make certain that rites are addressed so that the new rituals are reinforced.

• Cultural Infrastructure. If not addressed by the team, the various components of the cultural infrastructure can cause serious harm. By addressing this element as part of the design process, the team can mitigate this problem. The danger lies with the gossips, spies, and whisperers who fill their role in the infrastructure by spreading information. Because they are not an integral part of the change effort, their information is not complete and often wrong. They cause problems in the organization, resulting in a great deal of wasted time doing damage control. It is not possible to eliminate this problem because the people that fill these roles are engrained in the culture. However, it is possible to mitigate the problem by continuously communicating what is going on. Once people know what and why a change initiative is being developed, those in the infrastructure are less likely to be able to spread information that could harm the effort.

11.8 The Internal Consultant as the Team Catalyst

In many industries catalysts are added to the process in order to facilitate change and support creation of the final product. In the world of change, the internal consultant is the catalyst. The team is the group that the catalysts affect in order to bring change to the organization. Just as the catalyst injected into the production process does not become part of the finished product, you as the internal consultant need to take care that you serve the same role. You facilitate, focus, and help expedite the change being developed by the team. But they are the ones doing the work and delivering the end product.

Five Things to Think About or Do

1. Write out the definition of team in its component parts. Consider a team of which you have been a part and relate how these parts were or were not addressed. For the parts that were missed, what could you have done as the internal consultant to correct this problem?
2. Think about teams you have known, where the membership was less than 4 or greater than 12. What problems did they encounter? Were they able to deliver what they had been tasked to do?
3. Review the list of things that need to be addressed when selecting team members. Develop a checklist that you can use the next time you are involved in this process to assure the correct people are assigned.
4. Review the section that describes the internal consultant's role in the team process. For teams with which you have been involved, did you do the things listed? If not, why not? What will you do differently in the future?
5. Think about the role that the eight elements of change and the four elements of culture play in helping a team be successful. Identify a team that failed to deliver on their requirements. What elements were missing? Identify a team that was successful. Did they address all of the elements?

WORKING WITH
EXTERNAL CONSULTANTS

Internal and external consultants both can add value
The trick is knowing how to get both at the same time

12.1 The Problem

Imagine that the vice president of your company has contacted you about providing your internal consultant skills to help develop and implement a complete change to the maintenance work process in one of the plants within your company. This is a major undertaking. Furthermore, the company does not have all of the internal resources necessary to staff the effort nor the desired level of experience in the area of change management. As a result, you determine that external consulting support is needed. This is not a bad solution and is one that is employed by companies every day. However, it does add one additional element to the process redesign equation—that of balancing the work of external consultants with that of the internal work team. This task can be successfully accomplished, but great care needs to be taken along the way to assure that it contributes to the success of the effort.

In this example, suppose that the external consultants were selected by your company's president based on a long-term personal relationship between the president and the consultant company's owner. When they arrived on site, they made it clear to you and the work team that it was their responsibility to deliver a successful change initiative. Based on this statement, they took over the entire effort; they sought very little if any input from the internal resources that comprised the change team. It is true that external consultants can bring a great deal of experience to the change process. But ignoring the on-site personnel is detrimental to long-term sustainability.

The result of this approach was that the external consultant company drove the change effort, deployed it, and in the short term experienced a successful outcome. Eventually your company's president no longer felt that the external consultants were required. After all, the change initiative was a success. The site's maintenance process had been changed. So the external consultants left. Unfortunately, the change process that they had developed and were supporting immediately began to erode, eventually failing completely.

An alternate scenario again is one where the external consultants were hired based on a personal relationship with your company's president. This time when they arrived on site, it was immediately clear to you and the work team that they had no relevant maintenance or reliability experience. The result was that they took a subservient role to your team of internal resources and contributed very little. The problem here is that good external consultants can bring real value to a work initiative, but having consultants who are not equipped to do this costs you money and delivers little value. In addition, in order to sustain them as part of the overall effort, they require a lot of training and attention so that they can provide at least some benefit. If you don't do this, they will act as an anchor and hold back progress.

The next scenario is the ideal. In this case, the consultants come on board and immediately inform you and your team that they are here to support your effort, but not to lead it. They want to work as partners so that the best knowledge and experience of both the internal and external resources are brought to bear on delivering a successful change initiative. To make this an even better outcome, as the process is developed, they prove true to their word.

The first two of these scenarios are detrimental to any change initiative because the external consultants are not fulfilling the role they should play. It is only when they work in partnership with the internal consultants and the work team that maximum value can be achieved.

12.2 What Is a Change Management Consultant?

I have worked with external consultants for many years on various change-related initiatives. If you are like me, and are trying to make a step change or improvement, you will probably need their services. The trick is to pick the right consultants offering the most beneficial services. The right consultants will add value. The downside of selecting the wrong consultants is that they will set the effort back. Even worse, they could destroy what you have worked to achieve.

Change management consultants are individuals, groups, and sometimes even entire companies with an expert level of knowledge that they can use to support your change initiative. Their support includes:

- Assessing your current work process and identifying performance gaps
- Providing expert advice about what other companies, both in and outside your industry, are doing to improve
- Helping to develop a vision and a model of your work processes in the future
- Assisting in the preparation of a detailed work plan or roadmap for change
- Working to support the redesign of the work process in conjunction with personnel at the site
- Helping move the process forward in a planned and organized manner
- Supporting the leadership of the project
- Providing resources to support the work effort and to supplement site personnel
- Evaluating and helping to implement software that supports the process
- Developing and delivering training

Not every consultant can offer every one of these types of support. Nor will every consultant be an expert in your area of change. Therefore, you need to carefully evaluate consultants before you select one. You need to identify which services you need, then find one or more consultants who can provide these services. As you evaluate the consultants, consider the type of company they have, their areas of expertise, and their philosophy for conducting the work.

Type of Company.

At the small end, consultant firms can be individuals who are self employed and have expertise in a specific field, often in a specific industry. At the other end are large consulting firms with thousands of employees, addressing many areas of the work process. Some firms operate in a niche market, focusing their efforts in a key area. These firms are usually the small-to-medium sized firms, groups, or even individuals. Because of their lack of multiple areas of expertise and lack of depth of resources, these smaller companies are often somewhat limited in the size of the project they can undertake. However, they may be more cost-effective, depending on the size of your project, and offer more personalized service.

Areas of Expertise.

A consultant firm can not be everything to everyone. It usually has an area of expertise. That area may be a specific type of industry such as oil and gas, paper, or power. Their area of expertise may also be a specific function within the industry, such as maintenance, materials management, or finance. Single-industry specialists are fine if you can find one that has expertise specifically in your industry. Function-specific firms are also desirable if you only want to improve a specific function. A word of caution: Most work efforts that require consultants may start by addressing a specific function. However, if you really are redesigning a work process, these efforts can quickly become multifunctional. By limiting your effort to a specific functional area, you may quickly run into problems if the consultants are not capable of working beyond that area.

12.3 Work Philosophy

The work philosophy of change management consultants is also important. If their philosophy is not compatible with yours, then you may have serious problems.

The answers to three questions help determine your compatibility. First, what do the consultants want from the work effort? In other words, what will provide them satisfaction? Second, in the same vein, what do you want from the work effort? Third, what do you and the consultants want from your working relationship?

What Do the Consultants Want?

You may think that money is the only driving force for the majority of consultants. However, there are other factors. If consultants were solely interested in money, then some of their decisions and recommendations would never be made. Self-serving actions by consultants would make them more money, but the clients would not receive the optimum benefit. If this were the case, many clients would have a negative view of consultants, damaging their reputation and making it more difficult for them to get more business.

I think that the good consultants are motivated to improve the state of their clients' businesses and work processes. Their experience then expands their knowledge in the functional areas that they address. With these successes come referrals, an expanding business base, and ultimately revenue.

What Do You Want?

You have most likely sought a consultant for the following three reasons:

1. You want your change initiative to be successful.
2. You want success the first time. Management is often unforgiving when it comes to false starts. In addition, if you are not successful with change initiatives the first time, the organization may not give you a second chance.
3. You want to optimize the value of the change effort in the areas of quality, schedule, resources, and cost.

The drivers for these include:

1. You want to bring in additional experience to help drive the change initiative.
2. You don't have a broad base of functional knowledge other

than your own internal consultant experience or that of selected co-workers. You recognize outside help will be beneficial.

3. You are involved with other equally pressing internal consultant initiatives. Therefore, you do not have the time, nor are you provided the time, needed to manage this specific work process change over an extended period.

How do you reconcile your requirements in the first list with the drivers in the second? The answer is to hire consultants. They can bridge the gap between your business drivers and your requirements. By bridging that gap, they help you accomplish your goals.

What Do You Both Want?

Knowing what the consultants hope to achieve and having a clear understanding of what you want to achieve leads you to the task of finding consultants whose work philosophy matches yours. However, I am not referring to the consultants' knowledge or expertise as it relates to subject matter. Instead I am describing how you and the consultants want to approach the work. Your approaches must be in harmony for the effort to be successful. Specifically, you should look at two key elements: involvement and control. To what extent are the consultants involved in the project? To what degree do the consultants control the project? A bad mix of control and involvement by either party will be disastrous for the change effort. For example,

• Both the client and the consultants want the other party to control the project. As a result, there is minimal control. The project flounders and, in the end, fails.

• The client over-controls the consultants. As a result, the process yields little or no value. All of the consultants' valuable expertise is lost.

• The client abdicates control and is not involved. The consultants, by default, take over and run the project. As a result, the effort is successful only as long as the consultants are on-site. As soon as the consultants leave the effort fails. Why would you expect otherwise? The owners are no longer present.

You can think of the work philosophy as a control–involvement matrix. In Figure 12-1, the x-axis is the level of control and involvement that the consultants have over the project. The y-axis is the level of involvement and control that you (your company) have over the project.

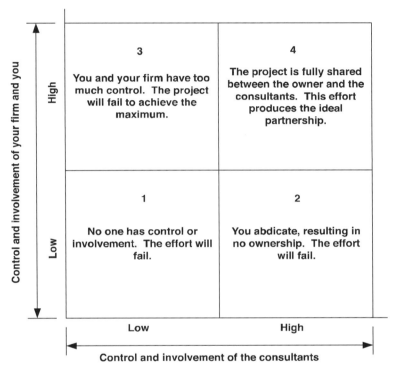

Figure 12-1 The Control / Involvement Matrix

Let's examine each of the quadrants in Figure 12-1 to gain a better understanding of the involvement–control relationship between you and your consultants.

Quadrant 1—Involvement and Control of Both You and the Consultant Are Low.

Good consultants do not actively seek low involvement and control. They are aware of the balance required between consultant and client for a successful effort. Consultants who end up with low involvement are often responding to clients who initially had good intentions when they hired the consultant, but then cut costs. What happens is the client begins to see that a successful change effort with the proper level of consultant involvement is expensive. In such cases, the client works with the consultants to reduce the cost. Because consultants basically sell their knowledge and skills, cost reduction usually leads to reduced scope and reduced resources. You hired the consultants for these specific attributes, yet, as the client, you have lowered their involvement and control. Additionally, you have dramatically lowered the value that they bring to the work effort.

At the same time, if the client doesn't recognize the significance of the change requirements on its resources and fails to appoint a full-time team, then it has further lowered the likelihood of success. With both client and consultants having low involvement, the project is doomed. If this approach is taken at the outset, or if you see this approach being applied as the project evolves, stop the project until you can correct the approach before the disaster strikes. Unless you correct this problem, the change effort will be less than successful. Even worse, the desired change may never take place. Additionally, the organization will have little or no tolerance for change in the future.

Quadrant 2—Involvement and Control Is High for the Consultant and Low for You

Here, the consultants have high involvement and control, whereas yours is low. This combination occurs when you and your company do not appreciate the significance of your role as an internal consultant in work process change. As a result, you are not assigned full time nor does your site management see the need for a full-time team, even though one is required. Assuming you haven't followed the Quadrant 1 approach, your company is willing to spend money on consultants, using them to fill the voids in your own change team. The good consultants will see the flaw in a Quadrant 2 approach (lower client involvement). They will probably even tell you about it. Having that information and acting on it, however, are two different things. If you still choose to let the consultants fill this void, they will. As a result, you abdicate your responsibility for both the effort and the outcome.

The real problem does not appear until the work effort has ended and the consultants leave. In the Quadrant 2 approach, you don't own the change process; the consultants do by default. When they leave your site, the change effort goes with them. Good consultants will try to prevent this problem by developing a sound transition plan that transfers ownership from them to you. However, you must be willing to accept ownership. The question becomes: If your firm didn't take ownership during the work, will they take it on at its conclusion?

Quadrant 3—Involvement and Control Is High for You and Low for the Consultant

In this scenario, the consultants have low involvement and control, whereas yours is high. When this combination occurs, your company has realized the need for a dedicated project team and the company team has accepted responsibility for the effort. This is good. The downside of this quadrant is that the company has allowed the consultants little involvement or control. This approach lowers the cost, but also lowers the resources and knowledge applied to the effort by the consultants. In most cases, this problem is caused by the client. However, the consultants may have overextended their resources, also leading to a resource and knowledge deficit. The result of this approach is that the change will have little, if any, effect on the existing process. With little consultant input, the company team will tend to replicate what is existing. At the minimum, the company will make only incremental changes when a major step change is required.

Quadrant 4—Involvement and Control Is High for You and the Consultant

The consultants and you both have high levels of involvement and control. This combination is the best of the four approaches. Your company has recognized the need for a dedicated team. You have realized that the correct amount of consultant time and effort, while not being inexpensive, adds value to your business. This quadrant represents the partnership approach. When the consultants are finished with this type of initiative, you have site ownership. What was created will be perpetuated.

The optimum *approach* to a consultant – client relationship is Quadrant 4. You and the consultants collaborate to achieve success for your firm, while allowing the consultants to achieve their objectives. The balance is difficult to achieve. For a change process to be successful, however, the balance is necessary. You need to be aware that it is your goal from the outset. All of your relations with the consultants—from initial selection, through the project, to the time they leave the site—must be driven with a Quadrant 4 approach.

12.4 Internal and External Consultants Each Have Value

Establishing a change initiative where you (as the internal consultant), the external consultants, and the work team all work effectively in Quadrant 4 is far easier said than done. As the site internal consultant, you need to support the site work team and not take ownership of the effort. The external consultants have the same responsibility. In addition, you and the external consultants need to work out a balance so that each of you contribute the individualized value that you bring to the initiative without one overshadowing the other and without completely taking control.

Figure 12-2 visually depicts this balancing act. Too much con-

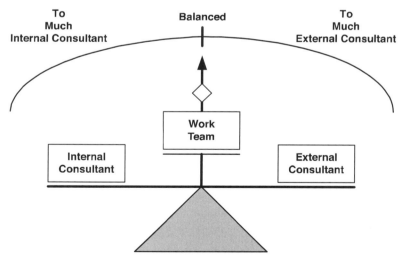

Figure 12-2 The Balancing Act

trol in either direction unbalances the work team; it forces them to lose the overall control and ownership that they should maintain.

But how can this be done? Don't the internal consultant and the external consultants each bring the same skill set to the effort and, as a result, have conflicts over who adds what value? The answer is no! Table 12-1 compares the differing values brought to the work team by the internal consultant and the external consultants respectively.

As you can see, the value that the external and internal consultants bring to the process are complementary. If addressed as such, they will certainly maximize the effort. However, there needs to be someone providing primary guidance. This can't be the work

Table 12-1 External / Internal Consultant Values

Category	External Consultant	Internal Consultant
Knowledge	Broad-based knowledge from working with numerous clients across various industries.	Deeper knowledge of the existing process, its history, and the specific difficulties that will be required to change it.
Involvement	The expectation is short-term involvement because no engagement lasts forever. Therefore, the desire is to deliver maximum value over the term of the engagement. Key areas include assessment, readiness, and deployment.	Recognition that the initiative will be long term and they will need to be around to help promote ownership and sustainability.
View	Multiple industry and numerous initiatives provide for a very broad-based view of the work.	Focused on the company due to the fact that they are a part of the organization. Possibly outside view from prior employers.
Culture	Understanding of culture and how to change it from exposure to multiple industries and work sites.	In-depth understanding of site's culture, who the key players are within the culture, and what is needed to alter it.
Leadership	Know-how regarding how to work effectively with organizational leaders on strategically-focused work initiatives.	Specific knowledge of existing site leadership, their issues, and what they want addressed. May not be as effective in dealing with the senior leadership because they are part of the organization.
The Eight Elements of Change and the Four Elements of Culture	General topical knowledge of how to weave them into a strategically-focused work initiative.	Specific site knowledge regarding each element.

team or the site because they are the ones being guided. It also can't be the external consultants because they will not be onsite forever. That leaves you, the internal consultant, in the role of providing overall guidance.

How you handle this role is critical to your success and the success of the effort. You need to maintain a position where both you and the external consultants work in a collaborative way. But it needs to be recognized that you are steering the ship. Figure 12-1 shows this balance as you try to work between Quadrants 2 and 3. Holding the external consultants in Quadrant 2 means that you lose oversight and the project will suffer when the external consultants leave. Forcing them to work in Quadrant 3 will lessen the value they can provide. Once again we arrive at the conclusion that the work needs to proceed from Quadrant 4.

You may be thinking that this is a conflict based on what I just said about you as the internal consultant leading the effort. It really boils down to working as equals, but you are slightly more equal that the external consultants. After all, you will still be there after they have gone onto other assignments. But how do you do this? The answer is by establishing clear roles and responsibilities for both yourself and the external consultant. Because there are many tasks associated with the work initiative, my suggestion is to sit down with the external consultants and establish a RACI Chart at the outset of the engagement. I have described the use and value of the RACI Chart in my book *Improving Maintenance and Reliability Through Cultural Change* and have included this section in Appendix 2. This may be a new experience for the external consultants, but it will go a long way to establishing who does what and maximizing the value that each of you can deliver to the overall effort.

12.5 How Clients Can Hinder the Process

Let's assume that you decide to use outside consultants for your change project, and have gone through the selection process. The consultants are now starting work. Another assumption has been made automatically: that the consultants can help. Under most circumstances, this is true. However, even the right consultants don't always help. The risk of failure—or a false start in a change effort—can be extremely dangerous to the organization. Efforts of this type that fail can seldom be restarted until years later. Most organizations simply cannot deal with change efforts too often.

Even if the project has been properly developed and the consultants have been carefully selected, what can get in the way of success? There are many factors including the following:

1. You expect too much too soon.
2. You do not provide sufficient information or communication.
3. You do not provide visible support to either the consultant or the process.
4. You do not have a vision, a goal, or a detailed roadmap for the effort.
5. You over control or under control the work.
6. You abdicate control or ownership.
7. You fail to provide for the needs of the organization, consultant, or the process.
8. You do not adapt as the changes evolves. Recall the discussion about spiral learning.
9. You overextend or under extend yourself.
10. You create adversarial relations.

As you were reading this list, you probably noticed that all of these factors were problems that the client caused. Can't the consultants cause problems as well? Yes, they can, as we will next discuss. However, they have experience in the change process. It is, after all, their business. The client on the other hand, has less experience and is often the prime contributor. There is no easy answer to this problem. As the internal consultant, you must pay close and ongoing attention to the effort as it evolves, identify problems that you are creating early, and correct them.

12.6 How Consultants Can Hinder the Process

In many cases, the consultants are the ones to cause a project to fail. Let's look at some of the ways that this can happen. Given this information, you can look for the symptoms and avert problems that could occur from lack of attention.

The consultants deviate from the original scope.

Some consultants try to generate more work or revenue for their firm. Be wary of consultants who constantly suggest additional work beyond the scope for which they were contracted. Don't confuse this problem with cases where the consultants legitimately identify work that may be out of the original scope, but is now needed.

If you and your consultants are operating from a Quadrant 4 position, then you should not have surprises of this sort. The additional scope would be recognized and addressed by the team. You can also avoid this problem by having a rigid scope control process. Then your response is that, while it would be nice to add whatever the consultant has suggested, the team needs to address the suggestion and the additional cost before approving it. The additional work can always be done later.

The consultants are over-controlling of the project.

Remembering Figure 12-1, over-control is Quadrant 3 behavior—the consultants' level of control is high and yours is low. If the consultants are pushing you into this quadrant, even though you initially started out as partners in Quadrant 4, then you need to have a serious heart-to-heart discussion with them and correct the problem. As you enter into the discussion, however, keep an open mind. The fault may not actually be their responsibility. It can also develop if your team has abdicated responsibility.

The consultants have the wrong people on the job.

This problem is a very real possibility. The consultants may have applied enough resources, but the wrong resources to the work. They may have misunderstood your requirements. In addition, the people they intended to assign to your project may have left the company or be tied up longer than expected with another client.

They may also be trying to work through a cost issue without upsetting you. This last problem is something you need to watch for, especially if you have pressured the consultant to spend less money. The level of resources they provide is based on your scope and what you are willing to pay. If you have a set budget and want specific resources, they will attempt to accommodate you. You may, however, be assigned resources that are not the right ones for your work effort. If this occurs, your team needs to review the requirements established at the beginning. Did you expect more than you paid for or have the consultants delivered less than agreed? In either case you need to correct the problem

The consultants have the wrong philosophy about how to run the project.
If you have done your homework and properly selected the consultants, this problem should never happen. At minimum, you should have the right consultants based on your preparatory work. If cutting corners at your end results in hiring the wrong consultants, you need to correct the problem sooner rather than later if you still want your project to have any opportunity for success

12.7 Working as Partners

Change management consultants can help your effort be successful. However, even with all that consultants bring to the effort, you must not abdicate your ownership of either the effort or the long-term outcome. You are there for the long haul. The external consultants are gone when the contract is over. Look at the consultants as partners, but you own a majority of the stock. If you approach your change effort in this manner, you will be successful.

12.8 What the Client Needs to Do

So far we have discussed internal consultants and external consultants. We've also looked at the role of the change team in forming a cohesive bond that allows each group to deliver value to the effort. We have not yet discussed the client. Often the client is the one that requests or even requires that external consultants be added. There are many reasons for this request, among which are:
1. They do not believe that the change can be handled by an internal consultant and want outside support.
2. They recognize a lack of the needed or correct internal resources.
3. They want change now and think external consultants can

deliver it.

4. They have prior involvement with external consultants and have seen the value that they can deliver.
5. They have been sold on the "magic pill" a quick fix solution to a long-term problem.

In any case, as the internal consultant, you are going to have to accomplish the change initiative with an "outsider" as part of the overall effort. As we have seen, if handled correctly, great value can be achieved. Conversely, if not handled correctly, great and long-lasting damage can be done. The key thing you must do is have the management team that brought the external consultants on board allow you to handle them as well as the work initiative. If this can be achieved and the management team can maintain their role as client, then you can merge both external and internal skills to deliver the value that you, the work team, and your site all seek.

Five Things to Think About or Do

1. Consider consultants who you have worked with where you felt that they worked with you as a partner. What was it about the type of company, their area of expertise, or their work philosophy that supported a partnership approach?
2. For the same consultants, review what it was that you wanted from them and what they wanted from you. How did these wants mesh? If there were areas where they did not, what did the two of you do to correct the lack of alignment?
3. Review the four quadrants shown in Figure 12-1. Identify consultant relationships you have had that fit into each one. What problems or benefits did working in these quadrants have for the work initiative?
4. Review the categories in Table 12-1. For each of the categories, provide examples where the work of external and / or internal consultants contributed value to work initiatives in which you were involved.
5. Build a RACI Chart for an existing or future consultant work effort. Develop a plan of how you would orchestrate this activity. For more details regarding RACI charts, see Appendix 2 page 303.

BUSINESS ETHICS FOR
INTERNAL CONSULTANTS

Internal consultants need to do things right
Even more important is that they do the right things

13.1 Do No Harm

Part of the Hippocratic Oath that doctors take as they enter the medical profession is "Do no harm." This same statement equally applies to the work done by internal consultants. The statement also applies at many different levels within internal consultants' area of responsibility. These areas include the company, the plant site, the department, the work group, and all the people who populate all of these groups. Additionally you do not want to do any harm to yourself so that you can have a long and productive career in the internal consultant field. The overall ethical principle of doing no harm can lead to some very interesting dilemmas:

•As the site internal consultant, you are supporting a change initiative that is failing. Much of the reason behind the failure was your inattention to detail as the initiative was developed and subsequently deployed. Because of this, the initiative failed before it even had a chance to get started. Management is trying to figure out why the effort failed so that they can institute some corrective action. You realize that if you keep your mouth shut, the team will be blamed and your credibility will not be damaged. What should you do?

• The change team with whom you are working has developed some very innovative ideas. Your manager doesn't realize that many of these ideas were developed independently of your involvement. He thinks that these were your ideas and is preparing to recommend you for a promotion. What do you do?

• You and the leaders of the change team you are supporting do not get along. You do not like their leadership style or the fact that they will take credit for the work of the team in order to promote their career. At the last meeting, the leaders got the team moving in a direction that you know from experience will cause serious problems for the team, but more specifically for the leaders. In fact, they may even be removed from the project. What do you do?

• You discover several errors in the training program being developed to introduce the new change initiative to the plant site. Corrections will take time and will delay that training class. It is important that the training be correct. In addition, the errors that currently exist will convey some incorrect information. To make matters worse, your bosses are bringing their managers to the first training class. You know that the managers won't recognize the errors and the students won't either until after deployment. What do you do?

• You have committed to come into work on Sunday to work with several members of the change team as they put together the finishing touches on the presentation which will explain the change initiative to site management. Your brother-in-law just called to tell you he has two tickets to the football game. What do you do?

• The change team that you are supporting is almost finished developing the details of their change initiative. Your manager calls and wants you to start immediately on another initiative out of town. When you explain your need to help the team finish the work, your manager becomes highly aggravated and tells you that it is a direct order to report to the new assignment Monday. You realize that if you do this, the current initiative will flounder and fall short of the expectations of the management team that brought you into the effort. What do you do?

• The change initiative on which you are working has immense potential benefits to your company. However, part of the change is the closing of two of your company's plants. You and the team recognize that this will put over one thousand people out of work. Although there are other solutions that have fewer benefits, they don't require plant closings and the associated layoffs. Remembering your commitment to do no harm, what do you do?

• The change initiative on which you are working is flawed. You recognize it and so does the team for which you are consulting. However, the site manager is putting extreme pressure on the team to implement it as soon as possible. The underlying reason is that the manager believes it will provide excellent short-term benefits. The manager, who is counting on getting a promotion before the inherent problems are identified, doesn't care about the site or the site personnel. As the internal consultant on the project, what do you do?

There is no simple answer to the question posed at the end of the listed ethical scenarios. Yet as an internal consultant, you are faced with ethical decisions of a greater or lesser degree each day. How you answer these questions—and how you act as a result of the answers you arrive at—will determine the success or failure of the initiative in which you are involved. Even more important, your answers to these questions and your subsequent actions dictate who you are as an internal consultant and as a person. They provide evidence for all to see as to whether or not you are ethical in how you work and how you conduct yourself on a daily basis.

There is another aspect of the ethical guideline associated with doing no harm. It is based on the fact that, no matter what outcome results from your internal consultant activities, it is going to be viewed by someone or some group outside of the initiative as doing harm to someone or some group. Unfortunately you can't avoid this fact because some of the outcomes from the initiatives on which you work may close plants, dissolve departments, or cause layoffs. The adage "to thy own self be true" applies here because you will never be able to please everyone. Furthermore, not everyone has the full understanding of the drivers that led you and the group you are supporting to the conclusions you have reached. Another way to view this is:

If the company survives, so does the plant.
If the plant survives, so do the departments.
If the departments survive, so do the work groups.

If the work groups survive, so do the people.

If you view your activities in this manner, the standards or ethical direction you provide to the change team will do harm, if harm has to be done, to the least number while simultaneously protecting the company.

13.2 Ethics Defined

What are ethics? If you were asked to define ethics specifically as they relate to an internal consultant, you might come up with many varied answers. Webster's Dictionary defines ethics as:

Rules or standards governing the conduct of a person or members of a profession

Thinking of ethics as a set of standards that govern a person's or group's conduct makes this concept slightly easier to understand. As an internal consultant, your role is to support a change team in moving the organization from a current state with which they are dissatisfied to some future desired end state. Ethics or the standards governing our behavior are the invisible boundaries in which we have agreed to function as we move from the "as is" to our created 'to be" model. The area outside of these boundaries constitutes non-standard or unethical behavior. Operating within the boundaries constitutes operating in an ethical manner. These boundaries can be set differently based on the work you are doing, but I believe that the governing standard should be "Do no harm."

Webster's definition of ethical conduct seems to distinguish between rules or standards of correct behavior for a person and those for a member of a profession. I believe that these standards are joined. I do not believe that it is entirely possible to act ethically as a person, but unethically in your job—in this case as an internal consultant. I believe that who you are and how you act in your everyday life is virtually the same as you would act in your business dealings. For example, would you:

• Spend time volunteering as a sports coach, but not spend time teaching an employee a needed skill.

• Refrain from telling harmful lies to your family, but do the exact opposite at work?

• Interact in a positive manner at home, but in the opposite manner on the job?

In almost every instance, the answer for ethical people would be a resounding no! Why? Because ethical behavior is not something you turn on or off; it is the way you try to behave in all instances.

13.3 Why Do We Make Unethical Choices?

In his article on business ethics, John Maxwell cites three reasons why people make unethical choices in business.

Ethics of Convenience

In this case, it is easier or more convenient to act in an unethical manner than to do the right thing. Recalling the example of the flawed training program, an internal consultant may find it easier to move forward even though the training program had problems than to stop the effort, lose valuable time, and rework the training. Ethics of convenience are even easier to justify if no one will notice the problem because, in a sense, you will be able "to get away with it." The problem with this approach is that, even if others do not know, you will.

To Win

Unethical behavior often can come into play if behaving in the unethical fashion will enable you to win, have a successful outcome for the initiative, gain a promotion, or defeat a rival, or in any other circumstance where personal or group success will be the reward and justification for the unethical act. Behaving in an unethical manner may help you win the day. But as an internal consultant, you are doing the work as part of a long-term career. Over the long term, you actually don't win at all. In the example where the team leader was taking the group in the wrong direction, allowing this error to continue may have resulted in the leader being removed from the group. This may be something that you may have wished for due to the problems you were having with them. You win, but do you really? A valuable resource would be lost and harm would have been done to the individual and to the group.

Situational Ethics

The third reason for unethical behavior is driven by the situation of the moment. Acting according to circumstances, where unethical behavior is the outcome, means that you are allowing your standards of ethical behavior to change based on specific situations in which you find yourself involved. For an internal consultant, this is very dangerous because over time it will destroy your credibility with the organization you are servicing. As an internal consultant, it is your job to set standards of behavior. People watch this closely; over time, they begin to understand the ethics associated with your role and your mandate to do no harm. How do you feel they would react if the ethical behavior you exhibited varied with the situation at hand?

13.4 Who We Can Harm

As internal consultants, there are many ways that we can cause harm. Our recommendations, our guidance and direction, and at times the leadership aspects of our position—if these are not provided with high ethical standards, we can cause immediate and often irreparable harm to those around us. Not only that, but in trying not to harm one part of the business, we may harm another without thinking. This makes working under the mandate of "Do no harm" extremely difficult; it places many dilemmas in our path and burdens on our shoulders as we apply your internal consultant skills.

Consider the issues associated with not doing harm:

The Company

Not all companies are so large that they can overcome a poorly designed and implemented change initiative. In these circumstances, the guidance that you provide, if applied poorly, could even put a firm out of business. Suppose that a small plant was having production problems. Their rotating equipment was not reliable and they were struggling to accomplish even the simplest reliability strategy. To try to overcome their problems, they had you act as the internal consultant to a group trying to resolve the issue. You know that for long-term success, a reliability program will provide a greater degree of benefit than simply developing a process to respond to the "emergency of the moment." However, the plant management team doesn't see it that way; they want a process that will make repairs quickly and get the equipment back into production. To follow the ethical principle of doing no harm at the company level, you must convince the management team that their approach will lead them down the path to ruin. Ultimately this may not be possible, but as the internal consultant you must do your utmost to try.

The Plant

Other companies are larger and have more than one production facility. As an internal consultant, you may work for one plant or you may work for them all. In either case, your ethical responsibility is to strive to do what is ethically correct at the plant level. There are various scenarios that can cause plant-level ethical dilemmas. Suppose your change team recommends a major reliability initiative that will cost a considerable amount of money; the plant will lose money over the short term. Further suppose that the management team wishes to avoid looking bad in the eyes of the senior executives. They object to the plan, even though they know that the long-term benefits would be significant. If you do not work with the team to promote the change, but instead simply go along with the management team, then you are failing to serve at the plant level.

The Department

Within each plant are many departments with varied roles and responsibilities. As the internal consultant, you may work with one or many departments on the various change initiatives to which you are assigned. Harm in this case is doing something that would advance one department at the expense of another, or not providing the proper level of consulting that would hurt the department you are consulting.

The Work Group

These groups are formed to work on the various strategic initiatives identified by management. Failure to deliver your best effort can lead your group to a wrong conclusion or to be responsible for a poor deployment. Either case would hurt the reputation of the group and of the members who were a part of it.

The People

Working day-to-day at all of the organizational levels we have discussed so far are people. They make up the work groups, departments, plants, and ultimately the company. Failure to conduct yourself in an ethical manner and failure to demand the same from those with whom you work hurts these groups. Worse than that, it hurts the people who are part of the groups and who are counting on you to help them deliver value.

13.5 Conflicting Ethics

It is also possible that your ethical standards and those of the group with whom you are consulting may be in conflict. Consider the following two scenarios.

• The change team with whom you are working has to develop a work process that will improve the reliability of the equipment in the plant. As part of the development of the plan of action, they decide to recommend that a new reliability team be formed. They also recommend that those currently handling this type of work in various departments not be part of the new group. The reasoning is that these will be highly visible assignments; they want the jobs for themselves. Your internal consultant business ethics are in conflict with this position. Not only does it do harm, but you know for a fact that the incumbents are some of the best reliability engineers in the company.

• A work effort in which you have been involved for over a year is reaching its conclusion. The team has done a very good job. Everyone expects that the results, when deployed, will add immense value to the business. In addition the changes, although they will realign several groups, will have no real negative effect on anyone. The group is very pleased that they have done no harm and achieved the benefits that were requested when the initiative was started.

However, there is one person with whom many in the organization have had problems with in the past. In fact, on many occasions, they have proven harmful to the business. This person is going to gain from the new direction and realignment being set for the maintenance department. The team supports the promotion that they will receive because they feel that they will gain politically. At a meeting with senior management, do you raise your concerns about this individual's ability to handle the new work?

Each of these examples shows conflict between the ethics of the group and your ethical standards as an internal consultant. These two ethical conflicts are depicted in Figure 13-1. This figure also addresses when your ethics match those of the team—positively or negatively.

The quad diagram in Figure 13-1 uses the x-axis to measure the ethics of the work group as positive or negative. The y-axis shows your own ethical standards, also as positive or negative. In this configuration, we can examine ethics both when they are in align-

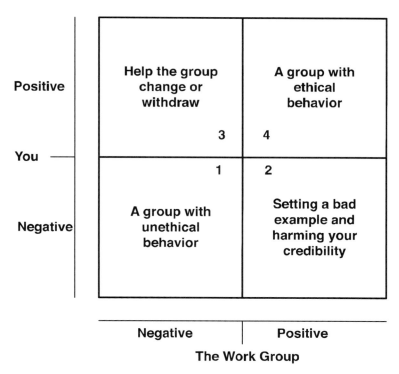

Figure 13-1 The Ethical Quad Diagram

ment and when they are in conflict. We can also evaluate what should be done in each case.

Quadrant 1—The entire team (including you) exhibit unethical behavior.

In this case, the work group will probably not last very long. Behavior of this type would become very clear to the client and result in the group being disbanded. Fortunately, this type of configuration, where both you and the work team exhibit unethical behavior, does not occur too often.

Quadrant 2—The work group has appropriate ethical standards, but you do not.

This type of scenario happens when the internal consultant is the one exhibiting unethical practices. As an internal consultant, behaving in this manner is not only damaging to the group and the group's performance; but also severely damaging to your credibility. Internal consultants behaving in this manner do not stay internal consultants for very long.

Quadrant 3—The work group acts unethically while you do not.

At times the work group—a team of people thrown together to develop and implement a change initiative—do not always know what ethical standards need to be applied to the effort. If you as the internal consultant behave in an ethical manner, you provide the group with a very positive example of the correct behavior. This gives them the opportunity to learn by example and change. It also promotes your credibility with the team.

However there are also times when the group will not change. This will cause a dilemma for you. You either start acting like the group, which will take you into Quad 1, or withdraw from the effort. The latter approach to this problem is far better because you prevent harming yourself. Eventually the group will be dissolved as a result of their actions.

Quadrant 4—Positive ethical alignment.

This is the ultimate goal. If achieved, it obviously will lead to a successful work and personal outcome for all involved.

13.6 Guidelines to Ethical Behavior

Ethical behavior is not something that you can turn on or turn off on demand. You need to live all phases of your life with the same set of ethical standards. As an internal consultant, you also have other ethical responsibilities as you go about facilitating work groups and providing consultation and direction to these groups. Things to consider as ethical guidelines in your internal consulting role as well as the rest of the time include:

1. Provide an ethical example to others.

As an internal consultant, it is up to you to provide an example of ethics in action to those around you. In your role, people see you as a subject matter expert when it comes to helping them develop strategies and tactics to implement change. In this process, you are also a role model for how to act. Behaving with high ethical standards will go a long way to assuring that those with whom you work behave in the same fashion. The evidence of this will be in the way that they take the work done by the team and deploy it throughout the plant site.

2. Expect ethical behavior from those with whom you work.

In addition to behaving ethically, you need to expect the same ethical standards from those with whom you are associated. In some cases, people may not even recognize the proper manner in which they should act. Suppose a member of the team is openly criticizing specific members of the management team. You must clearly state your expectations of ethical behavior in order for the team to correct this situation. Not only does it help people recognize their own ethical responsibilities, but it further serves to provide a positive role model of what is expected from the group and its members.

3. Don't confuse taking shortcuts with doing a proper job.

Often when pressure is at its highest level, people will cut corners. They rationalize that the schedule was achieved, even though the outcome may not be as good as it could have been. Cutting corners, as hard as you try to rationalize your behavior, delivers less than the best work of you and your team. Although the shortcut may have helped you to finish sooner, the failure to put forth your best effort will eventually be revealed. The lack of quality may not appear to your client until a much latter date; however you and those working closely with you will know it much sooner.

4. Recognize the work of others before that of yourself.

This is an important aspect of any form of consulting. Your job as an internal consultant is to support the success of others; not your own personal success. Therefore, you should not only recognize others before yourself, but go out of the way to assure that the team members are the ones to receive credit. Not only will this give them work satisfaction, but also go a long way to building a stronger sense of ownership that will live on long after you are no longer associated with the effort.

5. Teach what you know.

As an internal consultant, you possess a wealth of knowledge. However, if you keep it to yourself and fail to teach others what you know, the knowledge is wasted. Neither your associates nor the organization gains any benefit. One of your ethical standards, especially as an internal consultant, should be to impart knowledge all around you. In this manner, it will be used and potentially grow into something more value-added than even your imagined.

6. Don't assume anything.

Often you will experience actions by clients, people on the work team, or even those for whom the initiative was developed, that appear to be negative or even detrimental to the effort. Seeing this behavior often leads someone to draw conclusions and take action, only to recognize that the assumptions you made were in error. However, by this time, you may have said something or acted in some way that you immediately regret yet can not be undone. The way to avoid this is to get facts before taking action. Ethical behavior requires that this be your mode of operation. It is far easier to wait, gather the facts, and then act than to try to undo things that were said or said. Employing this standard in all of your actions also will rub off on those with whom you consult.

7. Treat everyone with dignity and respect.

I once worked for a manager who lived this ethical standard every day of his life. As a result of this behavior, he got far more in return. Those people who never seemed to get along with anyone behaved far differently in his presence and he got things accomplished with their help that others would have believed impossible. Treating people in this manner can go a long way to attaining a high degree of credibility and support in your role as an internal consultant.

8. Listen well.

One of the biggest problems many people have is that, as someone else is talking, they are formulating their response. Although they may have prepared a highly-intelligent response, when the other person is finished speaking they will have missed all of the important information that the person is providing. As an internal consultant, make listening well one of your key standards. To truly be able to do an effective job, you need all of the input you can get. If you listen well, you will get the information you need and more.

9. Build strong interrelationships.

As an internal consultant, the interrelationships you build, especially the good ones, are critical to your ultimate success. Your reputation as an internal consultant is built not only on the things you have done, but also, and more importantly, through the people you have supported and helped to achieve their personal and organizational goals. These interrelationships will provide you with both referrals for new work and a whole range of experts who you can contact for information to support future initiatives.

10. Be true to your word.

As an internal consultant your word is your bond. If you promise to do something, always make sure that you do it; where possible, exceed the expectations of the receiver. Your ethical standards for this element are very important. Groups count on their internal consultants to deliver on the promises that are made. Your reputation and credibility are at stake. Both are certainly enhanced if you deliver.

11. Communicate for complete understanding.

It is important for an internal consultant to make every effort to communicate effectively. Your work supports those who are developing and implementing new initiatives. Communicating the true meaning of what you are trying to achieve is a critical success factor. Through the feedback process, make certain that what you communicate is clearly understood. Failure to have this as part of your ethical standards can easily lead the best of efforts down the path to failure.

12. Serve the group.

A clear example of your ethics is how well you serve the group and your clients at all levels. Certainly in your role of internal consultant, you serve the group. But you need to do even more. No task should be too trivial or out of your scope, unless of course it is detrimental to the group, you client, or the goal you are seeking to achieve. This type of behavior elevates your position as an internal consultant. It is also a good example for other group members to emulate.

13. Teach others to lead.

It is true that you could lead every group with whom you work. However, that is not the role you should be playing when acting as an internal consultant. It is your job to help others to learn group leadership skills. Then they can be seen as the leaders and owners of the initiative on which they are working. Your role is to be in the background providing them with the needed skills and ultimately enabling them to succeed.

14. Promote ownership by the team.

You can not own the initiative or its outcome. If that were the case, then the effort would soon cease to exist once you left. Look to any initiative that is owned by the internal consultant and that is exactly what you will see. A good internal consultant needs to promote ownership by the team. Failure to do this is a violation of your ethical standards.

15. Seek feedback to improve.

In your role as an internal consultant, you are not perfect; there always is more to learn. One way to improve your own efforts is to ask for feedback from your peers, team members, clients, or anyone else who you believe can help identify areas where you can get better. Once you have this information, it is your job to do something with it. Put in place a corrective action program so that you can benefit from the feedback provided. Those who believe that they have no room for improvement will always fall short of the ultimate value that they can deliver.

16. Remember everyone makes mistakes.
No one is perfect. We all make mistakes, some minor but also some of a very major nature. Recognize this in yourself and in others. When mistakes occur, spend your time figuring out how to recover and learn from them. Don't spend your time blaming—this type of behavior serves no value and makes learning what really happened all the more difficult if not impossible.

17. Be a "can do" person.
There are two types of people, those who see opportunities at every turn and those who see the same opportunities as insurmountable obstacles. Your standard mode of operation as an internal consultant needs to be one where you act positively and see the possibilities vs. the barriers. This behavioral style energizes people and helps them overcome their own barriers to change .

18. Tell the truth.
Lying or even telling half truths always gets discovered. As an internal consultant, and even when you are not in your internal consultant role, you need to tell the truth at all times. Otherwise, you risk losing credibility and respect. Both are needed to be successful as an internal consultant or in every day life.

19. Praise openly; criticize in private.
People like to be praised for a job well done. They also like to receive this praise in front of their peers or work teams. It is a very positive form of recognition. This is something that you can do to help motivate others. On the other hand, if you have to criticize someone, do it in private. It allows them to discuss their issues and attempt to arrive at a corrective action plan. Handling criticism in

public is embarrassing; the person will not spend a great deal of time trying to identify areas for improvement. As the opportunity for praise or criticism arises, it is up as an internal consultant to you to handle it correctly.

20. Never say behind someone's back what you wouldn't say to their face.

What you say about people reflects on your ethical standards. If you are willing to make negative comments about someone behind their back, you will be looked upon as an untrustworthy person, willing to undermine a person's reputation indirectly. Instead, you should directly confront the individual about their behavior or performance, then help them address corrective action. In your internal consultant role, this approach will have a very positive connotation, saying to outsiders that your objective is to support people, not destroy them.

The bottom line for all of these suggestions is to behave ethically at all times, whether as an internal consultant or in life.

Five Things to Think About or Do

1. What does "do no harm" mean to you as it relates to internal consulting?
2. What are your ethical standards? Write them down along with real life examples where you have applied them.
3. Think of an unethical choice you have made in business. Why was it done? What was the outcome? Was there an ethical alternative and, if so, what was it?
4. Have you ever had an ethical conflict with a group with which you were working? How did you resolve it? Were you and the group satisfied with the outcome?
5. Review the 20 guidelines for ethical behavior. Which ones do you practice? For those that you don't practice, ask yourself why not?

COMPLETION OF THE WORK

When the internal consultant's work is done
The same should not be said for the work effort

14.1 Completion

It is true that a successful change initiative has no end. However, it is not true that the work of the internal consultants on that initiative must continue on forever. If this were the case, you would only get to work on one initiative for your entire career— this makes no sense. Therefore, at some point in time, the internal consultants' work needs to be completed; they need to move on, leaving behind a viable product that will add value to the business over the long term.

Consider a reliability change initiative designed to institute preventive maintenance practices into a plant site. You and the team you are supporting developed a detailed scope and, with a great deal of effort, have deployed a very comprehensive and apparently successful preventive maintenance process. When you and the team review the new work processes in place at the plant, you can clearly see that everyone recognizes the value of the preventive maintenance initiative. They are actively practicing the processes that they were taught in training.

From these observations you, your team, and the client (the plant manager) determine that deployment was successfully completed. In fact, the plant manager, who is very pleased with the outcome of the work effort, takes you and the team out to a very expensive dinner. At this point, it is time for you to move on to your next assignment and leave the operation of the preventive maintenance program to those working at the plant site.

Then about a month after you have left, you get a frantic call from the maintenance manager. The preventive maintenance compliance statistics, which were so good when you left, are dismal. The work crews assigned to the process are continually being moved to other jobs. Senior management is seriously questioning the value of the preventive maintenance effort. To make matters worse, one of the plant's major compressors just failed; the root cause was determined to be lack of preventive maintenance. The maintenance manager has already obtained approval from your boss for you to return and help get the program back on track. Therefore, you stop what you are doing and return to the plant.

When you show up Monday morning, the maintenance manager is there along with your former work team waiting for your arrival. As the meeting to discover the extent of the problem begins, it is clear to you that they all are waiting for you to tell them what to do to fix their problem.

There is something seriously wrong with this scenario. It does not take a lot of analysis for you to realize that, as an internal consultant, you failed to properly complete your assignment. As the work was progressing, you did not build a completion strategy into the work plan. Thus, the ownership was never transferred to the site personnel. When you left, so did the foundation that was supporting the preventive maintenance initiative. The result: it fell apart.

Failure to complete an internal consultant assignment in a definitive manner is a common mistake made by many internal consultants. For an initiative to be successful over the long term, this problem must be addressed. If you don't definitively end your assignment, you will find that you are continually sucked back in to solve problems that should be solved by the site owners. This is detrimental to those on site who should have the skills to keep the process viable. It is also detrimental to you, your ability to take on other initiatives, and your career.

14.2 Completion Defined

From an internal consultant's standpoint, project completion is:

That point after deployment when the internal consultant

can officially turn the initiative over to the site personnel and move on to their next assignment with a high degree of confidence that those on site will be able to sustain it.

As an internal consultant, you may be asked to return for a post audit. Still, you should never have to be recalled to get the initiative back on its feet or, even worse, to save it from extinction. If either of the two latter cases occurs, then you have not properly done your job as the internal consultant—you have not allowed the site personnel to take over management and ownership of the effort.

True completion for an internal consultant means that during the work initiative you assume a support, not a leadership, role. Therefore, as the effort draws to a close, the site personnel have taken on more and more ownership and are prepared to take it all as you exit.

A work initiative has three distinct parts: development, deployment, and post deployment.

Development

This is the phase where you and the work team you are supporting develop the initiative to the point that you are ready to deploy it. This includes developing details for work process, organizational structure, technology support, training, communication, rewards, and management approval. At the end of the development stage, all of the work to successfully implement the initiative should have been completed.

Deployment

This step is the actual roll-out of the initiative. At the end of the deployment stage, everything that was developed related to the initiative should be in place and properly functioning.

Post Deployment

After the initiative has been deployed, there is a period of time when people are becoming more and more skilled at making it function properly. As an internal consultant, you still are in a support role. But the site personnel should be running and fully responsible for the success of the initiative. At some point during the post deployment period, you can move on with full confidence that the work will continue.

This process is shown in Figure 14-1. The x-axis shows the three stages of a work initiative. Although the stages are shown as equal thirds of the x-axis, development and post deployment are in fact much larger, taking more time than the actual deployment phase. Development, depending on the initiative, could take a considerable amount of time. And until the initiative is ingrained in the work culture, post deployment could take years. The y-axis depicts percent ownership. As the initiative progresses, ownership is the sum of the ownership of the site and that of the internal consultant.

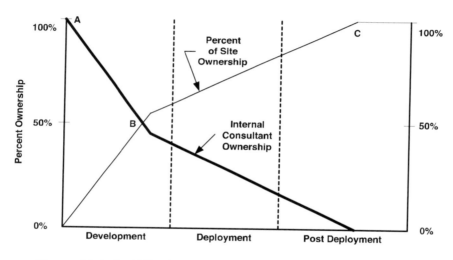

Figure 14-1 An Effective Site and Internal Consultant Process

At the outset of the initiative (point A), the internal consultant and the client hold 100% ownership. However, as the work team assembles and begins working through the development activities, a good internal consultant will begin the process of having the site personnel take on more and more ownership. Just before deployment, the site personnel should be the ones doing the leading and the internal consultant the supporting (point B). As deployment progresses, the site takes on more and more of the ownership and management of the effort. Then, at some point in the post deployment time period (point C), the internal consultant has transferred all responsibility for the success of the effort and is ready to leave. The most important point to be made in this diagram is that completion for the internal consultant begins during the development stage. Then as the effort progresses, more and more is turned over to the site personnel until (almost unknowingly) they own it all.

However, it doesn't always happen this way, as evidenced by the example at the beginning of this chapter. Often, the internal consultant doesn't provide the support and direction that enables the site personnel to ultimately assume full ownership. This type of flawed process is depicted in Figure 14-2.

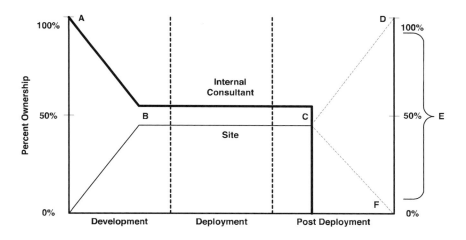

Figure 14-2 A Flawed Transition Process

In Figure 14-2, prior to deployment (point B), the internal consultant still has a great deal of ownership of the effort. The site personnel have not been developed so that they can begin to take on the majority role; therefore, they have not accepted it. As the deployment unfolds, the site still has not assumed the correct level of ownership. The internal consultant is essentially running the deployment process. In this manner, when the post deployment stage is reached and it is time for the internal consultant to move on (point C), the site personnel are not prepared and the initiative falters. There are three outcomes that can result following the departure of the internal consultant at point C.

- There is a brief lapse of ownership and the initiative initially suffers. However, the site personnel recognize and take on their responsibilities and eventually own the effort (point D).
- The site personnel are unable to take on 100% ownership and the process fails over time, eventually reaching point F where it no longer exists as a viable effort.
- The most likely alternative is somewhere between points D and F; labeled point E where there is more than 0% ownership but less than 100%. In these cases, depending on how much ownership is assumed by the site, the initiative will continue with varying degrees of success. However, if the internal consultant is not brought back in or if the site personnel never reach 100% ownership, the effort will fail over time.

It should be noted that the failures described are serious beyond the most obvious failure of a value-added site initiative. The deeper problem is that your credibility as an internal consultant will suffer because the failure of the effort will be attributed to the work you did with the site. In many cases, this could be the fault of the site personnel. But you need to share an equal portion of the blame because you never took the time to get the site ready to assume ownership.

A second serious problem is that those on site will become frustrated with the initiative as they try to make it succeed, but watch it disintegrate before their eyes. This cause great frustration within the organization. It also creates skepticism, making future change initiatives even harder to develop and successfully deploy.

14.3 Signs of Failure to Complete

As you support the transition of the site from dependence on you to one of initiative ownership, it is easy to get lost in the details. You may not even recognize at first that the effort is not on the correct track for a successful completion. Here are some signs that you are headed for problems. Notice that they reveal themselves during development and deployment, not post deployment.

If you fail to recognize these signs in the first two stages, then you are likely too late to do anything about taking the needed corrective actions.

- The vision is not clearly developed. This leaves the development team with a difficult problem. If they can not see a crystal clear picture of a future end state, then how can you expect them to develop a process to get the site involved or, for that matter, take on ownership of the effort?
- The vision is developed, but the goals are not being established. As we learned, the second part of the Goal Achievement Model, the goals, states broad-based objectives to support the achievement of the vision. A team that can not finalize the goals is not focused. It may not believe in what is trying to be accomplished. Without belief and commitment to the effort, ownership will never happen.
- The goals are established, but initiatives are not developed. Initiatives are steps designed to allow for the accomplishment the goals. A team that is not clear about the vision or the goals or doesn't believe in them will not be able to create initiatives. Once again, this failure will ultimately lead to lack of ownership and force you as the internal consultant to take on the role that the site is supposed to be acquiring.
- The initiatives are established, but no one is willing to develop or take ownership of the specific activities. In the Goal Achievement Model, activities are the actual steps identified to support the accomplishment of the initiatives. Because activities are very specific in nature and because you want ownership to be assumed by the site, the development team needs to assume responsibility for the accomplishment of the activities. Failure to take on this role is clear evidence of lack of commitment to accomplish the task. Moving into the development stage with this low level of involvement by the site will either cause the effort to fail in the deployment stage or prevent you from leaving as you assume the responsibilities for accomplishing them.
- You should also pay attention to activities that are being handled by the development team, but are not achieving the established milestones. Often when this occurs, the

internal consultant steps in and does the work to achieve the established milestone. This is often done because the internal consultant feels greatly responsible for the success of the effort. Although the milestone may be reached, it should be obvious that the internal consultant is setting in motion a process whereby the team will continually rely on the internal consultant to do the work, making eventual completion difficult, if not impossible.

There are other signs not associated with the Goal Achievement Model that also reveal long-term completion problems for you as the internal consultant.

- Leadership during both the development effort and the deployment phase is not being assumed by the site. The team plays a more supportive role, looking to you as the internal consultant to lead the effort.
- The team exhibits a great deal of frustration as they work their way through the development process. This is a normal event for teams that are forming, but it should not last over the long term. A team that is going to assume leadership and ownership will learn to function in a team-like manner.
- The interrelationships among people on the team are breaking down and can not seem to be patched up. This is often the result of frustration, which could have many and varied root causes. You must avoid stepping in and taking over in an effort to keep the team together and the effort on track. Such an action may correct the interrelationship issues for a short time. But ultimately you will have assumed ownership, something you do not want to do.
- Often the client, who is most likely not part of the development team, will get a sense that things are not working correctly. Their most frequent response is to ask you to make the needed corrections. Be careful! If you step in and correct the problem, you will invariably begin to take on the mantle of leader; your exit plan will be compromised.

14.4 What You Can Do to Assure Completion

Knowing all of the things to look out for—things that will position you as the leader and make your exit difficult—is fine. However, you must also know what you can do to correct the situation before it gets out of hand and whatever exit strategy you had in mind ceases to be viable.

From the time of entry, begin preparing yourself and the team for your exit. How you set the ground rules for your work and your interaction with the team will most certainly set the stage. It should also clearly establish what you will and will not do for the effort. In this manner, you are setting expectations that will define the team's role and your role in the process. In a sense, these expectations become a contract between you and the development team. Some of the things you should consider including are:

- I will help you understand the initiative as laid out in the scope of work so that the expectations of the client are fulfilled.
- I will provide guidance so that major deviations from the scope are avoided.
- Initially I will lead the team, but expect to turn over this role to the team members once the scope is clearly understood.
- I will help with team minutes, but they need to be the responsibility of a team member.
- I will initially handle logistics, but will turn this over to the team in short order.
- I will help develop presentations and reports. However, they are the status of the team's efforts; the team will be responsible for their development and, where necessary, for the presentation of the material.
- I will help create the training for the site personnel, but will not deliver it.
- I will monitor the progress of the team in achieving the initiative milestones and work with the team to develop corrective action where there is deviation. However it is the team's responsibility to get the initiative back on track if it slips.
- I will provide coaching and training to help the team succeed.

- I will go with you when you report to the client. You will present the report and I will support you.
- At the outset, based on the initiative's milestones, we will pick an exit date. We will collectively work towards my exit and the team's ultimate total ownership of the effort.

As you can see, there is a great deal that you can do in your internal consultant's role to support the initiative and those trying to develop it for deployment. However, if you closely review the items, you will recognize that they are designed to prepare the team and the site for total ownership and your departure.

14.5 The Completion Checklist

What are the indications that you are on the correct path to completion? Because there are so many aspects of a successfully completed change initiative, you need a checklist to be able to determine whether or not the initiative is positioning itself for completion and you for your exit. There are twelve elements that you must carefully consider when thinking about completion. Surprisingly enough these happen to be the eight elements of change and the four elements of culture.

Leadership and Organizational Values

For a change initiative to reach completion and move into the sustainability phase, the leadership must be heavily involved. This doesn't mean that they simply provide verbal support. If this is all that they do, the organization will quickly see that while they say they support the effort; actually they do not. What the leadership really needs to do is to embrace the change and be clear to the organization that work will now be done differently. This means that the change must become a part of the organization's culture. Then when events arise where the change needs to be applied, the organization will apply it without thinking.

An example of ingraining change in the culture is how various industries have addressed employee safety. An organization that is focused on safety will view every job as to whether it is safe to perform. If the job is not safe, the organization will make it safe before they begin the work. In past years, leadership may have preached safety. But many companies did not approach it as they do today. It was not part of their culture—that was the difference.

Work Process (including Rites and Rituals)

A change initiative can not be completed unless the work processes that are associated with it have been changed and are part of how business is performed. The second element of organizational culture, rituals and their supporting rites, addresses this area. The work process (rites) must be in place; the rituals that support the rites must be in place as well. Otherwise, the reinforcement will not be present; the site will quickly drift back into past processes

Structure and Role Models

The majority of change initiatives require that the structure be modified to support the work process. This is all well and good, but the role models must change as well. The new organizational structure is not sufficient. Those working within the new structure must model the behavior that is aligned with the process.

Group Learning

Course corrections will be needed as the process is deployed. The team needs to be able to recognize these needed corrections and make them without any involvement from you. They can't be so locked in to what they are deploying that they can't see or address the required correction. Your job prior to completion is to get them to understand this concept and quickly make the adjustments needed so the initiative can continue.

Technology

Almost everything we do in the reliability and maintenance arena involves computers and the applications they run to support the work. This also applies to change initiatives. Part of the completion process is making sure that the systems needed to support the work are present and functioning. For example, how successful do you think that a work planning change initiative would be without support software? The answer is not very successful; hence, technology needs to be part of your completion checklist.

Communication

In order to reach completion communications must be present on several levels.

- Communication to the site regarding the change and how it will affect them. This is an on-going effort and should start at the same time the initiative starts. The rule should be that more communications are better than fewer. People want to know what is going on; otherwise, they will speculate, usually with adverse results. Furthermore, upfront. continuous communication disables many parts of the cultural infrastructure that if left unattended could cause harm to the initiative. These parts include the gossips, spies, whisperers, keepers of the faith, and in some cases the story tellers. Refer to Chapter 5 for more details about the cultural infrastructure.
- Communications within the team so that everyone knows who is doing what and how the development and deployment is progressing. Because the tasks of the change team are being handled by many sub-groups or individuals, this communication is critical.
- Communications to the client or the site leadership so that they know what is going on and can lend their support as needed to help promote success.
- Communications in the form of feedback so that problems can be addressed and corrective action taken in a timely manner.

Interrelationships and the Cultural Infrastructure

Just as communications are essential, so are positive interrelationships among team members, work groups, and all of the individuals at the site. Without positive interrelationships, a successful change initiative will be difficult. Before you can reach completion, interrelationship problems need to be addressed and corrected.

Positive interrelationships also need to include the cultural infrastructure. However, you first need to know who are the members of this "behind the scenes" organization. This isn't as difficult as it may seem. Once they have been identified, you can bring them into the process. Having their support (or at least their understanding of your efforts) will minimize future problems.

Rewards

This is the reinforcement aspect of any change. For completion to be achieved, you need a way to sustain the effort. Rewards provide the means. Please note that genuinely successful long-term rewards seldom include money, which has a very short reinforcement life. The rewards that the team develops need to reinforce the work effort and be created in a way that motivates long-lasting culture change. Without properly designed rewards, there is a lack of reinforcement and completion. A sustainable initiative can not exist.

Preparing a Final Report

Once all of these pieces of the puzzle are in place, you should have a great degree of confidence that your exit will not be a traumatic event, but rather one that has been planned for from the start. There is still one more thing that you need to do—prepare a final report for the client. There needs to be a final document that, in a formal manner, marks the end of your involvement with the initiative. This document does not have to be lengthy, but should include the following parts;

- Purpose. Describe why you were brought in and what the overriding purpose was behind the initiative.
- Methodology. Identify the process that was employed to develop the initiative that is being deployed.
- The Goals for the Effort. This section explains in detail what the initiative is attempting to accomplish. An example is the creation of a proactive work process where only reactive work was present.
- Status to Date. This part describes the current status. It also provides background and the rationale as to why your work has been completed.
- Expected Benefits. This section lists the benefits that will be achieved over the short and long term. Where possible, this should be quantified and the numbers supported with factual information. However, this is not always possible, in which case more intangible benefits can be provided.
- Next Steps. Because change initiatives never really end, there are always next steps. One of these is the post-deployment audit.

14.6 Moving On But Checking Back

All throughout this chapter, I have described your goal from the outset as not only supporting the change initiative, but also building site ownership and, ultimately, your exit strategy. Exiting does not mean leaving forever. At some point in time, you and your client undoubtedly will want to review the progress made. You would also identify any corrective actions needed to get the effort back on track, if necessary, or to enhance the existing process.

This post-deployment audit should be conducted within six months of the deployment of the initiative. A shorter time period will not give the work process enough time to become ingrained. Waiting longer will allow the problems to become so ingrained that they may not be correctable.

You want to be part of the post-deployment audit. However, you do not want to lead it! If all of the work you did prior to this step was done correctly, you should not need to lead it nor should those on site want you to lead it. That doesn't mean you should not be part of it and provide the support, facilitation, and coaching that an internal consultant can provide.

There are many ways to conduct an audit. The most revealing is to interview people from a large cross section of the plant and find out what they really think about the effort and the benefits that were to be delivered. Interviews are not always easy. The amount of information you obtain will depend on how it is structured and the level of trust that the site personnel have with those doing the interviewing. This step is shown in Block #4 in Figure 14-3, following Steps 1–3, which were discussed earlier in this chapter.

The result of the auditing process will provide you and the audit team with findings block #5) both positive and negative. All findings should be communicated to the site (block #6) so that they recognize the good things that have been done and will see the negative ones corrected after analysis and corrective action plans are put into place. The negative ones are often factual and also often perceived. You need to reconvene the team (block #9) to sort out the truth from the perceived issues. Once accomplished you can conduct a change focused root cause failure analysis (block #10) and develop a corrective action plan (block #11). However you must let the team lead and own the analysis and the corrective

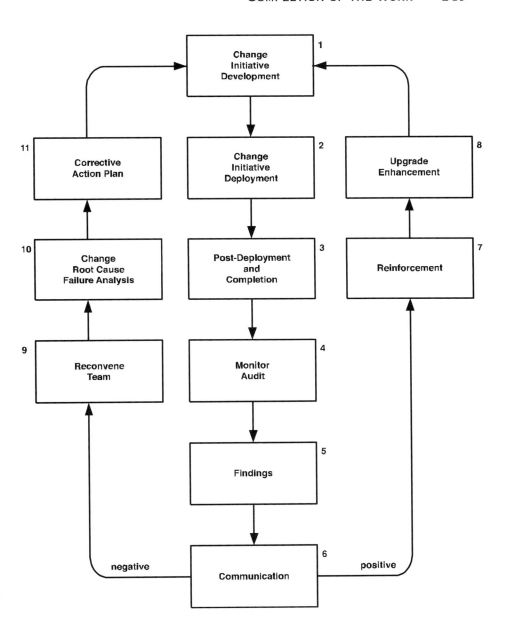

Figure 14-3 The Auditing Process

action. After all they are the ones who will have to implement the plan and make it a success.

As for the positive findings, they should be reinforced (block #7). As a result, there often will be ideas presented that could upgrade or enhance the effort. Here is an area where you can bring your internal consultant expertise to bear. Facilitating the process to improve upon the existing initiative is a good use of your internal consultant skills. Yet just like the negative pathway you should support the effort but not lead it.

14.7 One Last Note On Completion

If you and the team have done everything correctly, both the completion of your assignment and taking over full ownership of the initiative should be non-events. As you leave, note that your departure should not be the beginning of the end of the work, but rather the end of the beginning.

Five Things to Think About or Do

1. Consider change initiatives in which you have been involved. When did you start planning your exit? Based on this chapter, was it early enough in the process?
2. Review an initiative where the team never accepted full ownership and you were not able to move into a support role. Why did this happen and what could you have done differently to support team ownership?
3. Review the signs that indicate a failure of the completion process. Relate these signs of failure to initiatives on which you worked. Did these signs appear? If so, what did you do in the form of corrective action?
4. Develop an exit strategy template that can be used for current or future initiatives.
5. Develop a contract for future (or current) use indicating the things that you will do and the things that need to be done by the team. This can be included in your exit strategy template in item #4

READINESS AND SUSTAINABILITY

Without readiness there will be nothing to sustain
Without sustainability there is no need to get ready

15.1 The Need for Readiness

Imagine that the work team for which you are providing internal consultant support has developed a work process that the team as well as management believes will drastically improve the reliability of the plant equipment and the maintenance work process. After working on this initiative for many months, the deployment plans are drawn up and the date is set. Everyone on the project team is very excited about the prospects for significant improvement. They are looking forward to having the plant become a pacesetter within the company and maybe even within the industry. However, in the team's haste to deploy the results of their hard work, they neglected several things.

- The leadership in the field were never given the opportunity to learn about the revolutionizing changes as they were developed. Consequently, they never had to opportunity to buy in to them.
- The existing work process was never evaluated in detail so that transition plans could be developed to migrate from the "as is" process to the "to be" process dictated by the change.
- The organizational structure was not rebuilt to support the new initiative. The team erroneously believed that once the new process was implemented, the structure would adapt

to it without any major readjustment.
- The team failed to take into consideration past change efforts that failed so that they could make sure that the same mis takes were not made as they deployed the new process. They failed to learn.
- The existing computer systems were never evaluated to make certain that they could support the changes that were going to be implemented.
- The communication about the change was seriously lacking. Many key people who would be required to make the new initiative function properly were never trained. The organization had no idea of what was about to take place.
- It was recognized that the new process was going to have an effect on the interrelationships among departments, the work team, and individual members of the teams. But no plans were put in place to address this issue.
- The process required that people behave differently and, as a result, be rewarded differently for these new behaviors. However, altering the existing reward system was never considered.

All of these items are clear examples of the work team's failure to prepare the site for the change that was going to be implemented. The result? The site and its personnel were not ready, problems ensued, and the new change process failed. It didn't fail because the new process was flawed. It failed because the work team never took the time to consider what would be required to get the site and its personnel ready for the change. They didn't take change readiness into account. Their initiative as well as their company paid the price.

15.2 The Need for Sustainability

Now let us consider the other side of the process. Suppose all of the required readiness events were addressed and the problems cited never happened. Further suppose that the work team took all of the elements listed into consideration, addressed them all in a very proactive manner, and deployed the change initiative with great success. However, let us also suppose that as soon as the initiative was deployed, management returned all of the work team members back to the jobs they had prior to being assigned to the work initiative. The logic behind this action was that the initiative was a success. All of the change readiness activities had taken place and the deployment was successful. Based on this, the assumption was that the initiative was now going to be self sustaining.

Before very long, the errors associated with this assumption began to appear:

- New leaders were assigned to work at the plant. They had no way of learning or knowing about the former work initiative. As many new leaders do, they felt the need to impose their own work processes on the plant and show that they could make successful change in their new jobs.
- These new work processes were counter-productive to those put in place by the original change initiative. This conflict within two different work processes caused extreme confusion among the workforce. The result was reduced effectiveness and efficiency of the maintenance organization.
- The newly-imposed processes also were not compatible with the existing organizational structure; further confusion and often outright conflict ensued.
- Although members of the former work team tried to explain the value of the current process, no one was able to provide the learning they had acquired—the group was fragmented and no one had ownership of the overall effort.
- The computer systems used by maintenance, which had been adapted to the existing work process, didn't provide the correct information for the new one, resulting in even more confusion and frustration.
- The communication processes that had been carefully crafted as part of the former process broke down. The new managers felt that they were making improvements and did not feel the need to explain themselves.
- The interrelationships that had been established within the work team and across the site also dissolved. People struggled to understand why the existing process was being removed and why a new one, in which they had no input, was being implemented.

- The reward structure also broke down. Rewards are designed to support and reinforce action, but the actions expected under the new process were out of alignment with the current reward structure.

All of these items are clear examples of what can happen if a sound change initiative is put in place, but the site management fails to understand and develop a process to sustain it over the long haul. In our example, the plant encountered a severe reversal as the new managers implemented a change of their own. Without sustainability, the change doesn't have to be that severe to cause serious problems. Think about the results of not having a sustainability plan for plant sites that:

- Hire new people who do not understand the process or their role within it.
- Promote people to jobs where they need to understand how to function in new ways.
- Evolve and wish to include what they have learned into an upgraded process.
- Learn new ways to doing things that if added to the process would improve it even more.

All of these events require that a sustainability plan be in place if you wish your change initiative to survive and evolve.

15.3 The Need for Readiness and Sustainability

Failure to address both readiness and sustainability as two complementary parts of an overall change initiative places your effort on the road to failure. Both readiness and sustainability must work hand in hand so that you can deliver a successful effort. They also help assure you that all of the work you have done will be around for a long time to come. Consider each as a primary support for the change initiative, as depicted in Figure 15-1.

From the figure, it is easy to contemplate what will happen to the change if either of the supporting components are removed. This is something that we must make sure never happens.

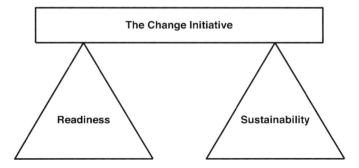

Figure 15-1 The Supports for Change

15.4 The Meaning of Change Readiness

We have shown a high-level example of the types of things that can happen if readiness is not part of the change equation, but what does readiness really mean to an organization that is about to put into place a change initiative?

Readiness for change is a state within an organization that is achieved when there is a high level of dissatisfaction with the old way of doing things, a clear vision of the future state and a defined set of next steps to get from where you are to where you want to go.

Anything less than readiness in all three components of the readiness equation will set you up for failure along the way.

- Lack of dissatisfaction with the current state will cause a high level of organizational discomfort if you try to make a change. Suppose the organization is highly reactive when it comes to how they handle their maintenance activities. Further suppose that they are very satisfied operating in this mode. How ready do you think they will be to change their process to one of planned and scheduled work?
- Lack of a vision will confuse the organization as you try to make changes. People don't typically like change, but they are often willing to go along with the effort if they can see where it is headed. A vision provides this; the lack of a long term vision makes it almost impossible for the majority to go along with the effort.
- Lack of next steps is even more destructive to successful

change implementation. Even if the organization can see the long term vision, they must be clear regarding the steps to get there. Without next steps, people will flounder. They will head off in the wrong direction and become frustrated and highly resistant.

The change readiness equation:

Successful Change Readiness = Dissatisfaction + Vision + Next Steps

---

This can also be shown as a set of steps that will enable you to vault the change hurdle.

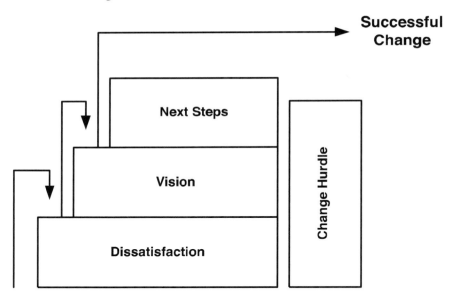

Figure 15-2 The Change Hurdle

15.5 Who Needs To Be Ready?

It is easy to think that readiness for change means that everyone in the organization needs to be ready for the change initiative to proceed. Aside from being impossible, this is not practical and not true. What needs to be in place within an organization is that a "critical mass" of those in the organization need to be ready for the change to take place.

What is a critical mass? It is not 100% of the organization. However, it is a sufficient number of people that, once the change has been initiated, will keep the effort progressing without outside or inside consultant support.

Think of the critical mass as a block positioned on a board held in the horizontal position. In this state, the block will not move without some external force being applied. However, as you raise one side of the board, you will eventually get to the angle where the block will move on its own and continue to move until it reaches the bottom. That is what critical mass is for the block. For the change initiative, it is that point where a sufficient number of people are engaged and believe in the change initiative so that it is self-perpetuating at the plant site.

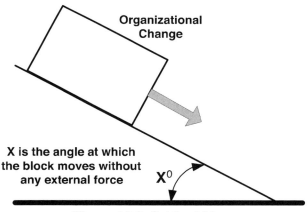

Figure 15-3 Critical Mass

How do you know when you reach this point? Although it is not easy to determine the exact point, some things are observable and will give you an indication. For example,

- People no longer consider the change as additional work. Instead they believe it is the work.
- People feel very uncomfortable when parts of the change process are not handled or completed properly. In fact, they may complain about the omission.

- People begin to suggest new "spin off" ideas that would make the initiative even better than when it was first deployed.
- You are not required to support the effort. In fact, when you observe others supporting themselves, it is a good indication that this has occurred.
- People who you knew were not bought in to the process now have become advocates and supporters.
- People who you knew were the resisters have at least become luke-warm defenders.
- If you tried to reinstitute the old way of doing things, people would be angry that you took away the tools that they needed to do their work.

If many of these changes in the behavior of the organization are taking place, then you will have a good indication that critical mass has been reached.

15.6 When Do You Get Ready?

The question is often asked by those engaged in the internal consulting process: when do you start the readiness process? The answer to this question is simple. You start the organization down the path of change readiness as soon as you decide that you are going to make a change. Readiness is not something that you add on just before you deploy the change. Readiness is something that begins when you begin. It then continues all of the way up to and through deployment.

An organization takes a long time to get ready, to accept that a new way of working is about to emerge in the plant, and to be prepared to adopt it. Readiness, therefore, is something that evolves through careful planning so that, when you deploy the changes developed, the organization is there along with you ready to make it a success. Those organizations that treat readiness as a step in the process and fail to initiate it at the outset will have an extremely difficult time to actually get the organization and the people that are part of it ready for the change.

15.7 How To Get Ready

You've now recognized who needs to be ready and that you need to start as soon as you decide that a change is in order. The next question is: How do you get ready? This is a multi-faceted effort. Review the example of section 15.1 that addresses the results of a work team that failed to prepare for change. Notice that the elements they missed are the eight elements of change—leadership, work process, structure, group learning, technology, communication, interrelationships, and rewards. These elements are also the eight elements that need to be addressed in any readiness plan. Additionally you must also take into consideration the four elements of culture—organizational values, role models, rites and rituals, and the cultural infrastructure—if true readiness is to be properly addressed.

When taken together, all of these elements will provide you with a well-developed readiness plan. But how do you combine all of them into the plan that needs to be developed? The answer is the Goal Achievement Model, modified to become the readiness plan that you need.

The Goal Achievement Model is a model that is used to take a vision and develop goals, initiatives, and activities that will support the accomplishment of the vision. For more details see my other two books: *Successfully Managing Change in Organizations: A Users Guide and Improving Maintenance and Reliability Through Cultural Change.* I have also provided details about the Goal Achievement Model in Appendix 1.

The same model that helps an organization attain its vision can help create the readiness plan. In this case the vision—the change effort—has already been established. The next level of the model, the creation of the goals, is also relatively simple because there is only one. Here, the goal is the development and employment of the readiness plan.

The next level of the model addresses the various initiatives needed to achieve the goal. This is where the eight elements of change enter the picture because we need a detailed initiative for each. Table 15-1 shows these three levels of the model and questions you need to ask to assure that this initiative is part of the readiness plan.

Table 15-1 The Initiative Portion of the Goal Achievement Model

Vision:
This is the change that is being established at the site.

Goal
Develop a readiness plan

Initiative Focus	**Initiative Questions**
Leadership as champions of the effort	Does the leadership support the change and does the site recognize their commitment to its success?
Work Process reconfigured if needed to support the change	Has the work process been reviewed so that it will positively support the change initiative?
Structure reconfigured if needed to support the change	Has the organizational structure been examined and, if needed, changed so that it will positively support the change initiative?
Group Learning • Learning from past mistakes • Developing and implementing comprehensive training	Have past change experiences been carefully examined? Have failures from the past been addressed so they will not be repeated? Has training been developed for the new initiative so that everyone will understand their roles and responsibilities? Has the training been developed so that it can be kept "ever green" as the process evolves?
Technology (existing) supports the change	Does the technology that is in place support the change initiative? If it will inhibit it, have plans been made to mitigate the problems?
Communication strategy developed and implemented to assure site understanding of the change	Has a detailed communication plan been developed and put into place so that the site will know what is going on and have a considerable amount of time for adjustment?
Interrelationships are maintained and efforts made to build strong new ones based on the change	Have the current interrelationships and the impact of the change been examined so that potential problems are recognized and can be addressed proactively?
Rewards in place to reinforce the change effort	Has the current reward structure been examined to determine if it is compatible with the change being implemented? If it is not, have new reward systems been developed?

With the initiatives established, it is time to move to the last element of the model: the activities. At this stage, you list each initiative you have developed, then determine in great detail what specific activities are required in order to have that initiative become successful. It is also at the activity level that you need to make certain the four elements of culture are being addressed.

For example, Table 15-2 shows activities associated with the initiative of leadership.

Table 15-2 Activities for the Initiative of Leadership
Initiative: Leadership
Associated Activities:
1. The site manager has accepted the role of change champion and has been briefed regarding the responsibilities of this role. 2. A process is in place to support the champion's role so that the site will recognize the importance of the change. 3. The site manager has established the change as part of the organizational values through word and deed. 4. All other members of the management team have been trained so that they understand the change and are committed to make it a success. 5. The change team has provided tools to support the management team as they become the role models for the change. 6. An audit process has been established so that the management team can quickly address their own performance gaps in supporting the change effort.

As you have probably recognized, the activities described under the initiative of leadership are not easy to accomplish nor are they one-time events. However, if you don't address the activity level for each of the initiatives in great detail, you will have failed to provide the readiness required for your change initiative to be successful.

15.8 Readiness and the Internal Consultant

As an internal consultant, you play a critical role in the readiness process. Typically change teams and individuals do not fully understand the complexity involved with "getting ready." You should teach them what is required and then guide them through the process so that the effort and their work will ultimately be successful. This means instructing them to use the Goal Achievement Model in the development of a detailed readiness plan.

Consider the example from the beginning of the chapter. How would an internal consultant help the change team get ready and avert failure?

- The leadership in the field were never given the opportunity to learn about the revolutionizing changes and never had to opportunity to buy in to them.
  - *As an internal consultant, your job is to make certain that the field leadership understands the change initiative from the outset and embraces it. This can be accomplished in many ways including change briefings or even short-duration training classes. Make sure that the change team under-stands how important buy-in and understanding are and that they proactively address these issues.*
- The existing work process was never evaluated in detail so that transition plans could be developed to migrate from the "as is" process to the "to be" process dictated by the change.
  - *The internal consultant should work with the change team to evaluate the "as is," develop the "to be," and identify the gap closure efforts required to make the change. Part of the internal consultant's work is supporting the development of the new process model. Even more important is a readiness plan to make the transition from the old process.*
- The organizational structure was not rebuilt to support the new initiative. The team erroneously believed that once the new process was implemented the structure would adapt to it without any major readjustment.
  - *Part of readiness is making certain that the change is supported by the structure. As an internal consultant, you should make certain this issue is addressed by the team.*

Often it takes considerable time to change an organization's struture. Because this may be a critical part of the readiness equation, you should make certain it is addressed in a timely fashion.

- The team failed to take into consideration past change efforts that failed so that they could avoid making the same mis takes.
  - ◆ *Group learning is an area where internal consultants have the advantage over external consultants. They have been at the site and know where the problems are from past change initiatives. You can bring this information to the team so that the mistakes of the past will not be repeated.*
  - ◆ *Good training is often missing in the learning part of change readiness. It is insufficient to conduct short classes and then expect people to understand and embrace the change. Learning needs to be much more than a single class. You can support the team as they build a sound education program around the change initiative.*
- The existing computer systems were never evaluated to make certain that they could support the changes that were going to be implemented.
  - ◆ *Technology is often overlooked because it is not something that can be easily replaced. However, there are ways that existing technology can be modified if needed to support a change. The internal consultant should make certain that the change team addresses this issue—not only related to how the technology works, but also related to the reports that are provided. These reports usually support the work process. If this is going to be changed, the internal consultant must make sure that the change team takes this into consideration.*
- The communication about the change was seriously lacking.
  - ◆ *This is one area where you can provide much needed help. There are many reasons why communication is not adequately provided related to change. It may be that the team feels it isn't needed, they are not ready to communicate, or the managment team wants to handle the process. In all cases, the internal consultant needs to make sure that a very detailed communictions plan is developed and employed. There is nothing worse than rolling out a change and having people resist because they*

> *didn't know about it. Your job is to make sure this never happens!*

- It was recognized that the new process was going to have an effect on the interrelationships, but no plans were put in place to address this issue.
  - ◆ *As we conduct our daily work, we build strong interrelationships with our work colleagues. Change often breaks these down, causing a high degree of organizational discomfort which often manifests itself as resistance. You need to make certain that the existing and new interrelationships are handled with care so that the change can proceed smoothly. Because this element is very intangible, the team may overlook it, but you can not.*
- The process required that people behave differently and, as a result, be rewarded differently for these new behaviors. However, altering the existing reward system was never considered.
  - ◆ *Rewards reinforce performance. The initiative will undoubtedly change what is considered good performance; in turn, the reward and reinforcement structure may need to be altered. For example, rewards for a reactive maintenance process would be far different than the rewards provided for reliability-focused work processes. Rewards also are usually established by organizations outside of the change team. Your job is to make sure that the change team brings the right people into their group when it becomes necessary to address the reward element of the readiness process. You should bring them in early too because rewards structures are often difficult and time consuming to change.*

Consider both the failed change process described at the beginning of this chapter and the things that you can and need to do as an internal consultant to reverse this process. You will quickly see the value that you can bring to the effort. In fact, an internal consultant who helps the group understand and employ a sound readiness plan may be the difference between success and failure.

15.9 Sustainability

Sustainability is at the opposite side of the change process. In the maintenance and reliability arena, the majority of work efforts

are considered as linear events. In other words, they have a beginning and an end. The project starts, it has a definite scope that is
executed, and it has a definite end point at which it is considered
complete. This is not the case with change initiatives! While it is
true that a change effort may have a beginning and a middle (execution), it has no end.

The difference between reliability / maintenance projects and
change initiatives is where the problem lies. At some point, a
change team will be disbanded; management will then expect the
initiative to be self-sustaining. If this took place after the change
was part of the organization's value system. all would be well, but
it usually doesn't. The typical approach is to disband the team several weeks after the change has been deployed. This is a serious
mistake. The result is that the change process often deteriorates
and, over a very short time, ceases to exist. This is not only a waste
of time for the site, the change team, and all of those affected. It is
an even bigger issue when you consider the value that has been
lost.

We develop a readiness plan at the outset of a change initiative. Similarly we need a sustainability plan so that six months
after deployment we won't wonder why the valuable change we
implemented has disappeared. For this reason alone, sustainability
is a critical part of any change.

> Sustainability can be defined as:
>> The ability of a change initiative to become integrated into
>> the organization's culture so that it isn't viewed as addi-
>> tional work, but rather it is viewed as how we work. Even
>> if there are changes within the leadership through promo-
>> tion, movement between sites, or new hires, continuity is
>> maintained.

For a change initiative to be sustained, the change must be
integrated into the culture. This is accomplished through the four
elements of culture. As with many of the readiness tasks, this concept is something that typically would not be considered by a
change team. It is, therefore, your responsibility as an internal consultant to make certain that they not only understand it, but also
incorporate it into the overall change process.

15.9.1 Organizational Values

Organizational values are the unwritten rules that govern our behavior at all times. Our actions are governed by these values. When behavioral decisions are needed, we don't have to think about what to do—the value system kicks in and we simply act to support the values. For example, a pump fails and the back-up is working perfectly. Do we 1) stop everything we are doing and make a quick-fix repair, or 2) take the time to analyze the problem and make a reliability repair that will eliminate all future failures of this nature? The maintenance organization will not have to think about the answer to this question because the organizational values will dictate the response.

The change that the team has developed and is deploying needs to become part of the organization's value system if it has any hope of being sustained over the long term. Obviously this will never be accomplished if the team is immediately disbanded after deployment because there is much still need to be done.

- The management team needs ongoing support to assure that they continuously reinforce the change. They have many things to think about and need a presence to make sure they do not forget the change that has just been implemented. This can be supported by the internal consultant, but needs to be owned by the site. The management team can do this at a high level. But someone from the change team needs to have this responsibility for a considerable time. Consequently, disbanding the team in it entirety should not be an option.

- Training needs to be an ongoing process. New people are hired, people change jobs, and still others take on new responsibilities. As a result, training must continue with someone responsible for the effort. Otherwise, people get on-the-job training by the incumbent. Over time, this causes a loss of a considerable amount of knowledge.

- The change initiative needs to be upgraded as the organization learns to employ it and arrives at new and better ways of making it work. With someone responsible for incorporating these additions to the process, it will be a dynamic tool. Without this level of support, it will become static and fail to deliver ongoing value. In addition, the ideas that are brought forward and not acted upon will create organizational frustration.

- The process needs to be audited so that emerging gaps and problems can be identified and addressed. Without an audit function, problems could be inherent in the process, yet never resolved, leading to frustration and resistance.

The above list is not all-inclusive but it does list the major topics. However, each site will have additional sustainability issues within the organizational value system that must be addressed if the change is to survive. As the internal consultant, you can support, coach, teach, and even lead people in the proper direction. But you can not own the effort—this must fall to the site personnel. To accomplish this, the management team must recognize that one or more individuals from the change team need to remain in this role.

15.9.2 Role Models

Role models are those people in our company or plant who people look to when they want to know what how to work and success looks like. If we want our change initiative to succeed, then our role models must be modeling the behavior that is aligned with the change. If they are not, they need coaching and training so that they can realign their work behaviors. Think about how difficult it would be to sustain a change initiative if the key people in the organization were modeling a behavior that was the opposite.

For example, a pump fails. As part of the current change initiative, engineering is supposed to conduct a root cause failure analysis. However the maintenance foreman, a recognized role model for the organization, never notifies engineering, makes a quick fix, and returns the pump to service. What do you think the message is that is sent to the organization? Do you think that if this continuously occurs they will believe that the change is actually a credible process?

A sustainability plan needs to require that the role models (current or future) understand that their performance is based on modeling the behavior supported by the change, and that they perform in this manner.

15.9.3 Rituals and the Supporting Rites

This element of an organization's culture refers to what we do (rituals) and how we reinforce what we do (rites). These two elements are usually an integral part of any change initiative. However, the pressing question is: How do we sustain the new process? Assuming that a sound readiness plan was utilized and the initiative has been put in place, we need a way to check on a continuous basis if the change is alive and well. This is handled by an internal audit of the process. Internal auditing will allow you to identify areas where there are gaps opening up between what you deployed as part of the change initiative and the current process (rituals and their supporting rites). Once these gaps are identified, gap closure strategies can then be employed to correct the problem areas.

Once again, you can recommend and support these audits. But the site needs to own both the effort and the associated corrective action needed to get the process back on track. This requirement clearly indicates that someone, preferably from the change team, has responsibility to initiate and conduct the audit. The management team has the responsibility to assure corrective action is taken. Failure to maintain the new set of rituals and their supporting rites will lead to unwanted results associated with your change initiative.

15.9.4 The Cultural Infrastructure

The cultural infrastructure is the hidden hierarchy of people and communication processes that binds the organization to the culture and sustains it over time. Key components include:

- Keepers of the Faith—individuals who act as mentors and explain how things are done within the context of the culture.
- Storytellers—individuals who reinforce the culture through stories of successes and failures so that everyone, by example, can see the cultures expected behavior.
- Gossips, Spies, Whisperers—individuals who pass information in various forms to the organization. They are important because they comprise the unofficial communications channels within the site.

Sustainability plans need to address all aspects of the cultural infrastructure. Plans need to be in place so that the correct behaviors are mentored. The stories that are told must reinforce the change. There needs to be a sufficiently high level of ongoing communication to weaken the "behind the scenes" problems that the gossips, spies, and whisperers can cause when communication is lacking.

15.10 Readiness and Sustainability: Two Equal Parts of the Process

As an internal consultant, you have two goals associated with a change initiative. First you need to help it become successful so that the organization and those within it can benefit from the new process. Second, you need to build your credibility. As an internal consultant that is what keeps you employed and gets the management team to request your services for new change initiatives. Both of these elements are intertwined. They are the result of successfully deploying change initiatives. Readiness and sustainability are two equal and important components of any change process. For your success and that of your company, they must be carefully considered and developed in detail. They must be part of each initiative in which you are involved.

Five Things To Think About or Do

1. After reading the chapter, write down why you believe readiness and sustainability are critical components of any successful change initiative.
2. Select a change initiative that succeeded. What readiness and sustainability plans were put in place to support the successful outcome?
3. Select a change initiative that failed. What readiness and sustainability components were missed? What would you now do differently? Do you think it could have resulted in a different outcome if readiness and sustainability were addressed?

4. Develop a PowerPoint presentation that explains the concepts of readiness and sustainability. Present it to management (your client) as well as to your work team.
5. For your next (or current) initiative, use the outline in Table 15-1 to develop a readiness Goal Achievement Model.

IMPROVING YOUR INTERNAL CONSULTING SKILLS

Positive feedback can make you feel good about your performance Negative feedback can make your performance even better

16.1 How Are You Doing?

Throughout this book we have discussed all aspects of the job of internal consulting. Individually and collectively, each of these is an important part of who you are and what you do in your internal consultant's role. However, as you take on initiatives, do the work, and bring the initiatives to conclusion, you need to find a way to provide yourself with 1) input as to how you are performing and 2) what specific areas need to be addressed. Why? So that over time and in future initiatives you can improve yourself and the work product that you deliver.

If you ask for input about your performance, your clients may or may not tell you. Those that don't know you can be very open and honest. They may provide you with valuable feedback, but this is often not the case. On the other hand, those that do know you and have worked with you may be reluctant to evaluate you. They may want to avoid telling you things that they perceive would hurt your feelings and, as a result, your relationship with them. Neither of these types of input is all that helpful. The former is usually provided without much long-term or detailed information about your performance. Yet the latter is watered-down and of little value towards long-term improvement.

Another type of feedback is at the project level. If you are given additional internal consulting tasks by the same managers time and again, you can reasonably assume that they feel you have done a good job. Otherwise, why would they have asked you back? However, if you are not given additional work, you may suspect that the value you delivered in your internal consultant role may not have been up to their expectations. This type of input to your performance is flawed. If you suspect that your past performance may be the reason you are not getting additional work, you may be making an erroneous judgment.

You need honest feedback to identify not just your strengths but also those areas where you can improve your work as an internal consultant. Therefore, what makes sense is to take the initiative. Find out the information you require to get better at your job. This proactive method is very important. By knowing your strengths, you can build on them. By knowing your areas for improvement, you can take proactive remedial action before any weaknesses become job or career threatening. The questions are: How do you do this? Once the information has been acquired and analyzed, how do you take the needed corrective action?

There are two avenues of pursuit for you to accomplish the goal of gaining remedial feedback. First is to ask for feedback from those with whom you worked including sponsors, clients, work groups, and other internal consultants. The second solution is to conduct a self assessment. Neither approach is mutually exclusive. Neither should be handled in a haphazard manner, if you really want to achieve meaningful results.

16.2 The Scope of the Feedback Effort

All of us have strengths and weaknesses. This is true of your work in the role of internal consultant. The trick is to identify both. Then you can strengthen those areas where you do well and improve those areas where you are weak. Chapters 4–13 cover what I consider to be the 12 Elements of Internal Consulting, one chapter dedicated to each. If you are going to survey your clients or do a self assessment, you should consider each of these elements both independently and collectively. The information you obtain will have great value.

This is only the first step. Once you have input, ask yourself, "Why do I (or my clients) feel this way about my performance?" Asking why is the standard root cause assessment approach. It is usually applied when we wish to determine why equipment failed, but it also works for assessments. Once you know and correct the root problem you improve your performance—or at least get started in the right direction. Conducting a feedback survey or a self assessment is a worthless exercise unless you do something with the information you acquired.

16.3 Client Feedback

Feedback from your clients at all levels is important to your success. If approached correctly, your clients will provide you with details regarding how and where you can improve. On the CD in the back of this book I have included what I believe is a good client survey. It has two parts.

The first part is for the client to answer five statements about your performance in each area of the 12 Elements of Internal Consulting. These questions are scored from 0–4, with 0 representing strong disagreement and 4 representing strong agreement. The result of this process will be a score from 0–20 for each of the elements.

Later (and also on the CD) these scores will appear on a web diagram that will graphically summarize the areas in which your clients believe you are a strong performer. The scores will also show areas where your client indicates you need to improve. Because you will most likely survey more than one person, the CD has been prepared for you to enter the score from multiple clients. You can then develop your web diagram as a result of all scores in a summarized fashion. See Section 16.5 for additional information related to the Internal Consultant web.

The second part of the survey is more subjective. For those questions where clients scored you 0 (strongly disagree) and 1 (disagree), clients are asked to explain why they score your performance in this manner. This input is valuable to your root cause assessment analysis; it provides you with the reasons behind the low scores.

The subjective section of the survey may provide you with input that is confusing or unclear. It would be easy to dismiss this feedback due to lack of clarity about its meaning. Don't take this course of action. Instead take the time to ask for clarification of any answers you don't understand. By asking why and seeking specific examples to validate their reasons, you will be able to 1) achieve clarity, 2) show the clients that you truly care about their input, 3) be able to clearly identify area for improvement, and 4) set up the initial level of your root cause failure analysis.

Be selective about who you ask for input and how many people you ask. You want to be sure that you find people who will:

- Take the time to properly complete the survey. After all, it covers 12 topics with 5 questions each for a total of 60 questions. It also seeks written input and examples that can take even more time.
- Know enough about the work you did to properly analyze your performance.
- Provide you with a cross-section of input from the level of client through the work group. You should also consider including those who are the recipients of the work product and who had interaction with you during its delivery.

With this said, it may not be possible for each person surveyed to address each area of your internal consultant performance. This is not a problem as long as you survey a sufficient number of people to generate a group of scores for each element.

16.4 Self Assessment

In addition to wanting feedback from your clients, you want feedback from yourself—a self assessment of your performance. I have found that I am my worst critic. My own feelings about my performance, strengths as well as weaknesses, are most often accurate. Where your clients may not recognize a weak area, you certainly do. An accurate self assessment will document them for you and enable you to build an objective plan for corrective action.

The format for the self assessment of your internal consultant skills is included on the CD at the back of the book. It uses questions similar to those asked of your clients except that the statements are worded for a self assessment. Each section has 5 questions that get scored 0 (strongly disagree), 1 (disagree), 2 (neutral), 3 (agree) or 4 (strongly agree). This survey will provide you with a score for each of the 12 Elements of Internal Consulting. Element scores that are from 0 to 3 are considered very poor, and those from 4 to 7 poor. Scores that are in these ranges are those where you need to do some self improvement.

Expect that your initial self assessment scores will be low. After all, even though you are most likely a good internal consultant, there is a great deal of material presented in this text that you may not have previously considered.

16.5 The Internal Consulting Web: Self Assessment View

Figure 16-1 is a sample of one of the elements in the Web of Internal Consulting spreadsheet on the CD. There are 12 of these, one for each of the 12 Elements of Internal Consulting. Each of these elements has the five statements related to your performance as an internal consultant as it pertains to that element. Select the score from 0 to 4 that most closely matches your opinion of what you believe your score should be. If you select a score of 0 or 1 for any statement, you need to consider some form or corrective action.

After you have completed your self assessment in each of the 12 categories, the scores will all automatically be transferred to the analytics section. Here you will find the total scores for each element on a table. Although the table has some value, the real value is in the web diagram shown in Figure 16-2.

As you can see, the web is an Excel radar diagram with 12 radial spokes. Each spoke represents one of the 12 Elements of Internal Consulting. The statement scores from the individual elements in the spreadsheet appear in the table and on the web. This diagram has immense value for you because it shows you your

		A. Strongly Agree (4 points)	B. Agree (3 points)	C. Neutral (2 points)	D. Disagree (1 point)	E. Strongly Disagree (0 points)
1	The internal consultant has made me aware early in the initiative the need for an exit strategy and I understand why.	●	○	○	○	○
2	In my dealings with the internal consultant, they have made it clear that one of their key completion roles is to assure that the site takes full ownership of what has been developed.	○	●	○	○	○
3	The internal consultant pays careful attention to the signs that indicate a potential failure of an initiative to complete. They have explained the rational for this to me.	○	●	○	○	○
4	The internal consultant has made me aware when signs indicating a potential failure to complete are present and have provided corrective action plans.	○	○	●	○	○
5	The internal consultant has created a contract with the change team so that they understand what they need to do to take on full ownership.	○	○	●	○	○
	Score>>>>>>>>>>>>>>	14				

Figure 16-1 One of the Elements on the Self Assessment Survey CD

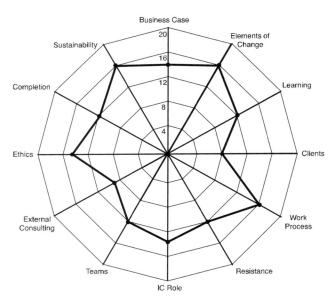

Figure 16-2 Your Web of Internal Consulting

perceived strengths and areas for improvement on the same diagram. Because the elements are both independent and collectively important, you can draw performance conclusions, then develop corrective actions plans in a far better manner than if you were only able to examine the scores in tabular form.

Suppose your scores for the 12 Elements of Internal Consulting were those shown in Figure 16-3.

Element	Score
Business Case	10
Elements of Change	11
Learning	8
Clients	5
Work Process	3
Resistance	9
IC Role	11
Teams	12
External Consulting	9
Ethics	11
Completion	8
Sustainability	10

Figure 16-3 An Example in Tabular Format

From these scores you would probably focus your attention on the several elements that scored lowest: business case, elements of change, clients, and sustainability. A web diagram has the potential for altering your perspective by showing the scores as a collective grouping. The table from Figure 16-3 is shown as a web diagram in Figure 16-4.

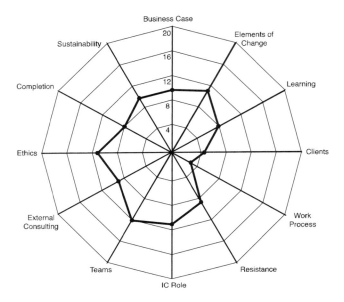

Figure 16-4 An Example Web

Unlike the table, the web shows a somewhat different picture of your self assessment data. The same four elements are still showing as low scores where remedial action is needed. However, the recognition that these elements are not mutually exclusive of the others and seeing their scores in this manner could lead you to conclude that you do not need to work on these three individually. Instead, you need to address them as part of a collective whole.

16.6 The Internal Consulting Web: Client's View

Earlier in this chapter we discussed the need to obtain client assessment information from more than one person. This input will provide you with a broad perspective of your performance from those for whom you have worked. Once you have received the surveys back, you need to enter the scores into the section on the spreadsheet for client input. This section has a simple spreadsheet with each category and the statements included in that category listed by number. There is a series of 10 columns so that you can input data from 10 client surveys. This section of the spreadsheet

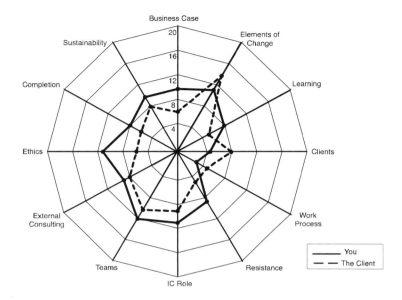

Figure 16-5 Your Scores and Those of Your Client

will summarize the statement scores and provide an average score for each question as well as an average score for each of the 12 Elements of Internal Consulting.

In addition, this section will generate a client web diagram which is identical to that shown in Figure 16-4 with one important exception—the information was generated by your customers! A second web diagram will also be generated showing your self-assessment scores overlaid with those from the client. An example of this is shown in Figure 16-5. In this diagram, the client scores show as a dashed line. On the CD, these web diagrams will appear as two different colors.

There is tremendous value in being able to view your self assessment web overlaid with that of you clients. It will show you three very important things about your performance:

1. Areas of strength where you and your client agree
2. Areas for improvement where you and your client agree
3. Areas where you and your client do not agree about your performance.

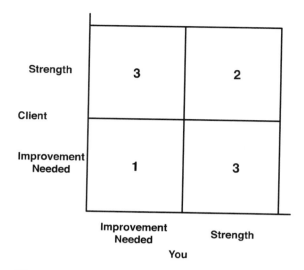

Figure 16-6 Three Perspectives on Performance

These three conditions are depicted in Figure 16-6 and identified with numbers that correspond to the next list.

In the block labeled 2, you and your client agree on your strengths. These should be maintained and even improved upon because you both recognize that these areas are your strong points. The fact that your client recognizes this as a strength is good reinforcement for you. Conversely, in the block labeled 1, you both feel that the element under review need some work so that you can get better. It is up to you to do something about this. At least the client has reinforced or identified an improvement area for you to address.

The blocks labeled 3 are where you need to do some review both retrospectively and with your clients. In the block in the upper left quadrant, your client has identified a specific element as a strength, but you have identified it as an area for improvement. Similarly in the lower right quadrant, your client has identified an element as needing improvement, yet you feel that it is one of your strengths.

You need to resolve these discrepancies. That is why I have provided space in the client survey for client-written examples. Using these statements about your performance can give you a sense of why a client may have evaluated your work differently than you evaluated yourself. Be aware that people are willing to check off scores based on their perception of your performance, yet may be reluctant to cite specific examples. Nevertheless, you need to resolve the differences. If your clients have not provided examples to support the lower scores, you may need to go and ask them.

16.7 What Do You Do With the Survey Information?

The surveys provide you with information from your client and your own self assessment about where the strengths and areas for improvement are in conflict. This conflict needs resolution, as do the elements where both you and your client agree that

Element	Chapter	Client Scores	Your Scores	Delta Scores	Delta > 8	Scores Both < 8
Business Case	4	7	8	1		
Elements	5	9	8	1		
Learning	6	10	10	0		
Clients	7	6	7	1		XX
Work Process	8	14	13	1		
Resistance	9	14	14	0		
Internal Consultant Role	10	8	10	2		
Teams	11	17	13	4		
External Consultants	12	11	11	0		
Ethics	13	16	10	6		
Completion	14	16	12	4		
Sustainability	15	13	4	9		XX

Figure 16-7 Comparison of Your and Your Client Scores

improvement is needed.

To resolve these discrepancies in perceived performance, you first need to prioritize them. Then work on the ones where major differences in the scores or total agreement in low scores exist. Minor differences can be excluded because a difference in a score of 8 or less could be attributed to work perceptions. It is the larger differences that need work.

The best way to handle this is to create a table for you and your client's scores as shown in Figure 16-7. You can do this manually or if you are using the CD allow it to be done for you in the spreadsheet.

In the fifth column labeled "Delta >8," you identify those elements where the difference in your score and your client's score is greater than 8. The higher the difference is, the greater is the potential problem. In Figure 16-7 the element that fits this category is sustainability. There also are potential problems where both you and the client score an element less than 8, even though the difference between your scores may be less than 8. In this example, the element that you and your clients have scored in this manner; is identified in the column labeled "Score < 8." If more than one element fits either the "Delta >8" or "Score <8" columns prioritize them from the highest to the lowest. These elements are the primary ones you will need to address in your corrective action plan.

16.8 Internal Consulting Performance Improvement

Now that the potential areas for improvement have been identified, you need to determine why they were identified as such and get to the root cause. As with all root cause analysis processes, identifying and addressing the root cause of the problem will go a long way to problem resolution.

- You can begin the process by selecting one of the elements on the list. Then ask yourself the following questions; one of these will apply to the element under review.
- Why did my client score my performance so low while I thought I did well?
- Why did I score myself so low while my client thought that I did well?

- Why did my client and I both score the element so low?

Let us examine the low sustainability score (you scored this a 4, your client scored this a 13, with a difference of 9). In this example, you would ask yourself question #2; it is the question that applies to this situation. Brainstorm some answers. Assume that by asking yourself this question you arrived at the following possible reasons:

- I did not have a site person identified as the initiative leader and did nothing to correct the situation.
- I took on too much and did not enable the site team to own the initiative.
- I didn't fully understand sustainability as it applies to this initiative and, hence, never considered it.
- I was pressured to get the initiative implemented and did not take the time to address sustainability.

These reasons to your "why" question are revealing, but need further review and personal clarification to determine if they correct.

After some thought, you are able to rule out bullets 1 and 3, but have discovered something about 2 and 4. They are linked! The client did pressure you to deliver on an unrealistic schedule. In turn, this pressure caused you to take on too much because the site personnel were busy with their regular jobs. As a result of the accelerated schedule, you never really had time to consider sustainability.

Because this element is one where you and the client differed, you also need to investigate why they scored you high vs. your low score. You have the client survey results. Therefore, you can check what was written. You find that they did not give any consideration to sustainability and considered the delivery on schedule to be sufficient. Hence, they scored you high on this element.

For the analysis of other elements that had score differences greater than 8 or low scores both below 8, this same approach should be applied. It may require that you drive deeper into the reasons why or go interview clients. In any case, the information you acquire will help you improve; that makes it worth the effort.

There are so many variations on what you will uncover, that

there is no prescription for what to do related to remedial action on your part. In some analyses, the area for you to improve may be obvious and easy to correct. In others, it may require outside reading or even a training class. In still others, you may be fine; the problem may reside with the client. Even if the problem is with the client, you are still the one who needs to do some corrective action, but in this case it is corrective action for others.

Five Things to Think About or Do

1. Try out the Web of Internal Consulting CD at the rear of this book so that you understand how it works. You may want to create a web with fictitious scores to test the process.
2. If you have done special projects or are an internal consultant, select a recently completed initiative and conduct an actual self assessment. Take a look at the web and identify areas for improvement.
3. Take the same initiative and have those who were you clients assess your performance. Enter this information into the spreadsheet and create a client's web.
4. Print out the web with both your and your client's scores, then examine it in detail. Create a table for yourself like that in Figure 16-7. Identify the areas where you and your client disagreed on an element by more than 8 points. Identify areas where you both scored below 8 for an element.
5. Select an element where improvement is needed and conduct a root cause analysis.

THE END OF THE BEGINNING

While this is the end of the book
I hope it is the beginning for those who have read it

17.1 The Guide

A wagon train heads west. It is going to be a long and difficult journey fraught with danger all along the way. However the people on the wagon train are not fearful of the unknown because that have experienced guides who know the way, know the dangers, and more than anything else know how to avoid the problems that could threaten their success. After many months of travel, they safely reach their destination. All throughout their travels, they relied heavily on their guides even for their very lives. But now they have arrived at their destination and are ready to settle the land and reap the benefits for which they have struggled so hard. This is the point at which the guides step aside. They has served their purpose; it is time for them to move on to the next wagon train and the next group of settlers who will be under their care.

Of course you are not literally going to have to cross swollen rivers, deal with blinding dust storms, fight off hostile enemies, endure physical hardships, and risk your very lives. Nevertheless, you are still on a significant journey.

The internal consultant's role is not unlike that of the wagon train guide. It is your job to guide those under your care safely from where they are to where they wish to be. This is not an easy task. There are pitfalls along the way and they take many forms, both obvious and not-so-obvious. However, once this task is accomplished, you need to step aside, fade into the background.

You must allow those you have guided to take charge of the journey from here and create the success they have worked so hard to achieve.

That is what this book has been about. It has been developed to help you hone your skills as an experienced guide so that you can help those placed under your care to change and improve how they do work.

17.2 The Summary

We started with a description of what an internal consultant's role is within the organization. A very important first step is understanding that, above all else, you are an internal consultant, regardless of whether you are called a project manager, a special projects coordinator, or any other organizationally-created title. Then we differentiated between strategic and tactical work, explaining why they can not co-exist for a person at the same time. This is also an important concept because, as an internal consultant, you will need to work in both arenas. But you need to understand clearly why you can't work in both at the same time.

Next we developed a business case for the internal consultant role. This is important for your organization, but even more so for you. Recognition by the organization of your true role and the value you can deliver is important, not only to help you deliver value to your organization, but also—and equally—to yourself.

Following this discussion, we looked at two critical components of the change process: 1) the soft skills or the Eight Elements of Change and 2) the organizational culture identified by the Four Elements of Culture. Companies initiate change all of the time. The concern is that many end in failure. Some fail so badly that the organization is highly skeptical of future change, making new initiatives difficult to enact for years to come. The reason for this failure is that change in these companies is only addressed at the hard skill level—initiatives such as improved planning, instituting preventive maintenance and the like. What these companies fail to take into consideration are the soft skills and the organizational culture. If these factors are out of the equation, then failure of the change initiative is not just likely; it is pre-ordained.

We also discussed one of the central elements of change: Group Learning. This element deserves a chapter unto itself. How and what an organization learns and how it feeds this learning back into the overall process of change is critical to its long-term success, and maybe even its very survival.

Then we moved onto a very important topic one that addressed your clients, how to work with them, and how to satisfy their expectations so that you are able to deliver value and get the next assignment. As an internal consultant, you are only as good as the last assignment. You need to do good work so that you get additional assignments.

The next three chapters dealt with the internal consultant's role, the process that needs to be followed for a successful outcome, and a very important chapter on resistance. Although many think that resistance needs to be overcome, that is far from the truth. Resistance needs to be understood and addressed. Those who are resisting are doing so for a reason. Change can only be successful if these reasons are understood and if the associated problems, perceived or real, are addressed.

The next two chapters considered two opposite ends of the work process spectrum: the work team at the site and the external consultant firm that is often brought in (whether you want them or not) to help you achieve the goals sought by senior management. Working effectively with both of these groups is important. After all, you are only the guide for the work team. You need to know how to guide them properly. External consultant are also guides, but like the wagon master, you need to utilize their expertise yet still maintain your overall guidance of the effort.

Chapter 13 addressed itself to business ethics for the internal consultant. Ethical behavior is most likely something you do in your internal consultant role without thinking. The purpose of this chapter was to get you thinking about this topic and, as a result, improving your focus in this area.

The next two chapters dealt with closure. First we discussed how to close out an initiative. Second we looked at how to support the initiative and sustain the changes that you and the team have delivered to the work site. Remember: change is not a project. It is a process that, once started, goes on and on as you learn and apply new things along the way toward your ultimate vision.

Finally Chapter 16 introduced the Internal Consultant's Web of Change. This web helps you measure not only the areas where you have strengths, but also the areas where you can improve. Even the best guides learn and apply new skills each time they go to work. That is how they continually improve and continue to add greater and greater value. Use the web to identify your own areas for improvement, and to solicit feedback from your clients and work teams. At times you may feel you have a strength where others may feel you need some improvement. Take this feedback as constructive; it can only make you better.

All of this information was designed to provide you with additional internal consulting tools and ideas to help you improve yourself. Applying these tools will enable you to add more value to your organization.

You may be asking yourself, "Where do I start?" My suggestion is to set your personal vision as becoming the best internal consultant that you can possibly be, then build a Goal Achievement Model from there. In this way, you can establish goals, initiatives, and specific activities to help guide you along the path to your vision. This process can help; I promise.

17.3 Final Thoughts

Last but not least, contact me through Industrial Press, Inc., or through the link to my website from within the Web of Internal Consultant Change with any comments or questions you may have. I promise I will get back to you promptly. Just as constructive feedback will help you get better, the same applies to the work that I do.

I would like to close with one final thought. Think about it because this is what a good internal consultant delivers to their organization time and time again.

Successful internal consulting

Is like casting a stone into a still pond

Initially ripples spread out in every direction

But quickly the pond returns to its state of clam perfection

But if you take note you will see

There has been a subtle change
While on the surface the pond is calm
Within it has been rearranged

Good luck in your role as an internal consultant and your efforts to add value to your company, your plant, your peers, and most of all yourself.

Steve Thomas

APPENDIX 1

THE GOAL ACHIEVEMENT MODEL

Appendix 1 has been included to provide details related to the Goal Achievement Model, a tool that can be used by the internal consultant in order to help clients achieve their vision. The information provided was extracted from my first book, Successfully Managing Change in Organizations: A Users Guide.

A1.1 The Goal Achievement Model

The Goal Achievement Model is a tool that links the goals, initiatives, and activities of the organization, department, team, and individuals to the overall vision. It makes these links in a way that all can easily see how their efforts contribute to the end result. The model also provides a global view—a way to see how efforts by one group can affect or be affected by the efforts of others.

Goal achievement overcomes many of the problems associated with achieving an organization's desired "to be" end state. It provides a clear understanding of the vision and shows how goals, initiatives, and specific activities performed by the site personnel link to it. Goal achievement requires focus as well as coordination among departments and groups. By the very nature of this coordination, goal achievement provides feedback to the participants, relating what they are doing to the achievement of the vision.

The Goal Achievement Model (Figure A-1) illustrates the method. Before looking specifically at how the model works, let's first clearly define its key components: the vision, goals, initiatives, activities, and measures.

Vision.

The vision is a single broad statement that describes the overall mission for the company, plant, or department. The vision is

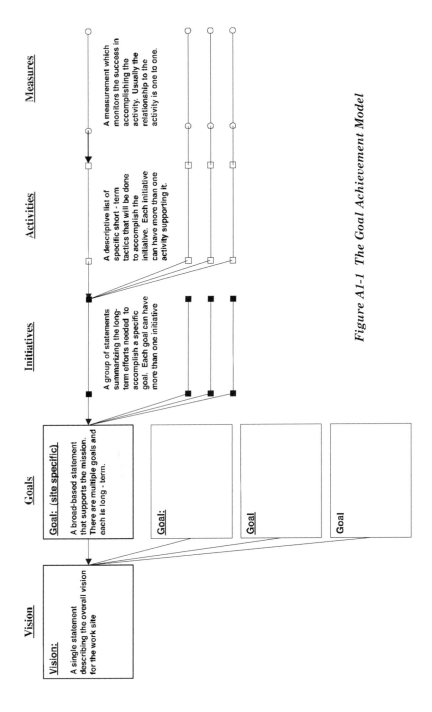

Figure A1-1 The Goal Achievement Model

stated in a way that employees can readily understand its relevance to them. It is aimed at a high-level purpose. For example, the mission might be "to operate the facility in a reliable manner so that the requirements of the customers are always satisfied."

Goals.
Goals are broad-based statements that support the mission in long-term, but specific ways. Typically, there will be several goals in support of the mission, with each one addressing a different aspect. Continuing with the vision described above, the goals could include:
1. Develop a comprehensive reliability program.
2. Improve the level of workforce skills.
3. Train the workforce to make decisions focused on reliability of the equipment and the processes.

These goals actually help to sharpen the vision, focusing on the word reliable.

Initiatives.
Initiatives are statements describing long-term efforts that will be made to accomplish a specific goal. Typically each goal will have several initiatives associated. Many different efforts are usually needed to accomplish a stated goal. Initiatives are generated by the groups who do the actual work. Including them in a visual model allows the various groups to see what others are doing. The model also provides a mechanism to avoid duplicated work or efforts that are counter productive. For the goal of developing a comprehensive reliability program, specific initiatives could include:
1. Establishing a program for predictive maintenance.
2. Establishing a program for preventive maintenance.
3. Developing a tracking tool so that work can be scheduled.

Notice the increasingly specific nature of these statements as we go from vision through goals and initiatives.

Activities.
Activities are the short-term, specific tactics that explain exactly and in detail what the group or individuals will do to accom-

plish each initiative. At this level, specific work steps and tasks are described. The activities are usually developed by the group responsible for carrying them out. Looking at the initiative of establishing a program for preventive maintenance, activities could include:

1. Determining which types of equipment should receive pre ventive maintenance.
2. Identifying the equipment and gathering data to load into the database.
3. Determining how to staff this activity.
4. Establishing how often each piece of equipment should receive preventive maintenance.
5. Developing a schedule for the work.
6. Developing a plan to monitor the completion of all of the tasks and to enforce the maintenance schedule.

As you can see, activities are short-term specific steps that you take to complete the initiative. Responsibility can be assigned to an individual or group for completion of the task. Without assigned responsibility, the effort gets lost in the everyday work.

Measures.
Measurement is the last, but perhaps the most vital component of the process. It tracks progress for each initiative and activity. Measures show if the work is on track; they hold everyone account-able.

A1.2 How the Goal Achievement Model Works

The Goal Achievement Model allows for a vision and high-level goals to be clearly stated and then converted from an abstract concept into something that can be clearly understood. This con-version is achieved through the continual development of the vision into successive levels of detail through the model. The vision of "operating the facility in a reliable manner" is abstract; it has various meanings for different groups and people within the com-pany. If we did not proceed further into the model, leaving people to create goals on this statement alone, the result would probably not be what was wanted. However, as you look at the subsequent levels and examples, you can see that the approach becomes very specific. The abstract nature disappears.

Each successive level becomes more detailed so that by the time you reach the Activity level, the work to be done is easily translated into actual work tasks. The model enables you to identify specific tasks that, when completed, support the vision. At the lower levels of the model, this is clearly achieved

Because the work done at the Activity level is recognizable in its contribution to the overall vision, employees can see the value of their efforts. Figure A-2 illustrates how each level of the model supports the next higher level. For example, the activity of gathering the data supports the initiative of developing the preventive maintenance program. In turn, this initiative supports the goal of creating the comprehensive reliability program that ultimately

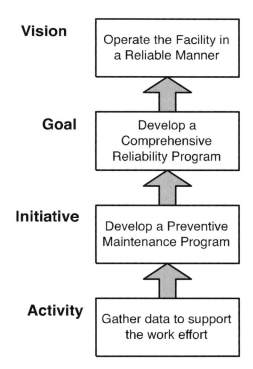

Figure A1-2 The Upward Relationship Between the Components

supports the overall mission. This connection is clear not only to the executive level but also to all who are involved. Thus, the model can help drive the success of the process.

Because groups within the facility can see what others are doing, they can eliminate conflicting activities or even those activities that negatively impact each other. The majority of companies today do not have an overabundance of resources. Therefore, you want to be sure that you and your group are working on the right things. When everyone's efforts are shown, the model allows each group to focus on the work that adds value for the company.

The measures established to track the activities provide evidence that progress is actually being made. How many times have you prepared goals, only to have them wind up in a desk drawer for future retrieval when you are asked, "How are you doing with your goals?" You find yourself scrambling to see what you have done, trying to make your accomplishments fit what you said you were going to do. Proper measures avoid this problem. If you report your measures on a regular basis, the chances that you will be scrambling at the end of the year are minimal.

A1.3   Goal Achievement: An Example

Let's now consider a full-scale example, one that focuses on an area of importance for all companies, especially those with production facilities. Suppose you are working in a plant that does not have a good safety record. Management decides it wants the plant to focus on improving its safety record, including the overall safety of the personnel and assets. The Goal Achievement Model is put into action. Therefore, plant management develops a vision for the site that states, "Provide a safe work environment." For a site that does not already have a good safety record, this vision is impossible to achieve in a short time. Using the Goal Achievement Model over time, however, will enable the organization to focus on all the components that will help it to succeed.

The vision can have many associated goals. This vision has several, identified in Figure A-3. The first, which will be further developed in this example, is to create an education program that raises the level of safety awareness in the plant. A second goal, to identify and correct unsafe acts, focuses on employee actions whereas a third goal, to identify and correct unsafe conditions, focuses on equipment and procedures. The last goal, to comply with safety-related legal requirements, looks at the plant in a broader environment. Note that the goals are still somewhat abstract, yet they have begun to sharpen the focus on the type of work that will be needed.

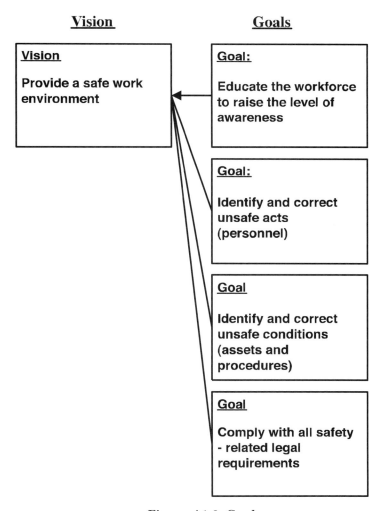

Figure A1-3 Goals

The next step, setting the initiatives, gets more specific, though it retains a long-term perspective. These steps should be done at the various organizational levels in the plant. Those who have to carry out initiatives and convert them into activities should have some ownership over the process. If the staff sets the initiatives, then it will be difficult for the employees at lower levels in the organization to feel that ownership. (The exception to this rule is for any initiative that may be specifically required at the corporate

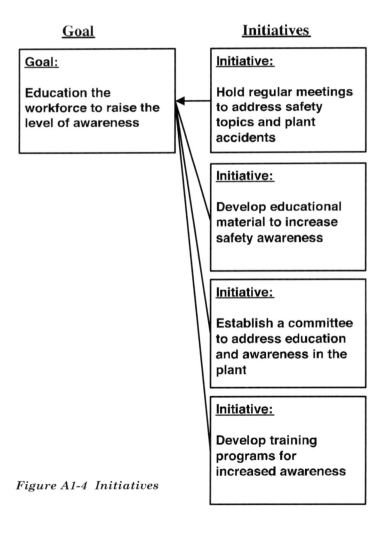

Figure A1-4 Initiatives

level. In this case, the staff sets the initiative and mid-level manage-ment must make it successful.)

In our example, the goals shown in Figure A-3 were presented to the plant personnel in order to increase their understanding and support. They formed teams to identify the initiatives needed to achieve these goals. Figure A-4 shows the initiatives developed for the first goal to "educate the workforce to raise the level of aware-ness."

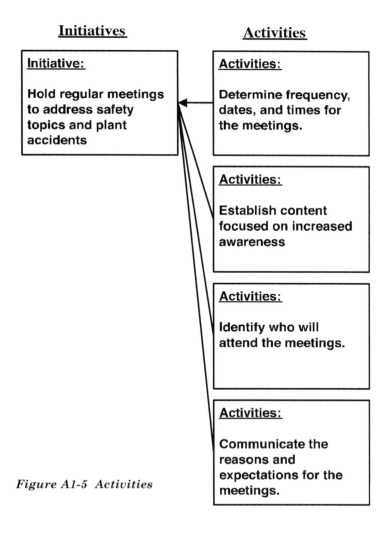

Figure A1-5 Activities

These are examples of initiatives within the goal of creating safety awareness. One of the better ways of generating this list is through group brainstorming, and then ranking the list by priority. Note that this goal has several initiatives, as would the other goals.

With initiatives established, the next step involves determining what activities (tactics) need to be developed. Figure A-5 zooms in on the activities identified to support the initiative: "Hold regular safety meetings to address safety topics and plant accidents."

The trick is not just to generate this list of activities, but also to establish a set of measures that track their progress. To complete the example, I selected the activity: "Determine frequency dates, and times for the meetings." Figure A-6 shows these measures that go with the activity.

To summarize this example, Figure A-7 provides a full diagram of all of the steps we have discussed. Take some time and see how the various steps flow from the original mission. Note how

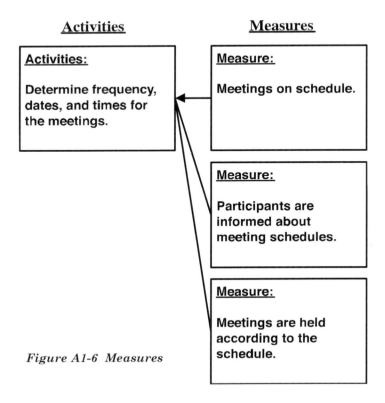

Figure A1-6 Measures

Vision	Goals	Initiatives	Activities	Measures
Vision — Provide a safe work environment	**Goal:** Educate the workforce to raise the level of awareness	**Initiative:** Hold regular meetings to address safety topics and plant accidents	**Activities:** Determine frequency, dates, and times for the meetings.	**Measure:** Meetings on schedule.
	Goal: Identify and correct unsafe acts (personnel)	**Initiative:** Develop educational material to increase safety awareness	**Activities:** Establish content focused on increased awareness	**Measure:** Participants are informed about meeting schedules.
	Goal Identify and correct unsafe conditions (assets and procedures)	**Initiative:** Establish a committee to address education and awareness in the plant	**Activities:** Identify who will attend the meetings.	**Measure:** Meetings are held according to the schedule.
	Goal Comply with all safety - related legal requirements	**Initiative:** Develop training programs for increased awareness	**Activities:** Communicate the reasons and expectations for the meetings.	

the accomplishment of a single activity can be tracked back so that those doing the work can clearly see how they are contributing.

APPENDIX 2

RACI

A2.1 RACI Charting

Creating a RACI diagram is a technique for identifying key work activities and functional areas of responsibility associated with a specific task in a matrix format. By representing the work task activities in this fashion, the roles and responsibilities of the functions associated with each task can be clearly delineated.

This approach will work well if you are trying to clearly identify the work distribution between you, the work team, the client, and the external consultant. Although it takes time to build, it clearly displays the roles and responsibilities of those involved in the initiative. It can also be used to bring people back on track if you see roles and responsibilities beginning to conflict as the effort progresses.

A2.2 RACI Guidelines

RACI stands for responsible, accountable, consulted, and informed; the four levels of involvement that the various job roles have as a part of a change initiative or for that matter any work activity. The details associated with each are as follows:

- Accountable. The position within the organization with sole (go / no go) authority over the specific task that is to be completed.
- Responsible. These are the job roles that are working on the task. Responsibility can be shared among several roles or it may even be assigned to an "A" role.
- Consulted. These job roles are consulted prior to the final

decision or action. This requires two-way communications.
- Informed. These job roles are informed about the decision or action. This does not have to be prior to the decision being made.

There are some general rules for creating RACI diagrams as follows:

1. Conduct the analysis with the group who will be working on the change initiative. In this way, they can discuss their roles and responsibilities as the chart is developed and ask for clarity at the outset.
2. Take time to do it right. Building a RACI Chart will take time, but it will be time well spent as the work effort unfolds.
3. Place accountability (A) and responsibility (R) where they truly belong. Think about who the final decision maker is for each task (A) and who is responsible to actually do it (R).
4. There must be only one "A" per activity—the job role solely accountable for the task. An "A" can also have "R" functions.
5. There can be multiple "R" roles for each task; however, too many Rs can cause problems.
6. Authority and accountability for action MUST be linked.
7. Assign consulting (C) and informing (I) roles with great care. Not every task requires them.

A2.3 The Job Roles

There can be many roles on a RACI Chart just as there can be a great many tasks. For the work we are doing related to site change initiatives, there are essentially five basic roles:
- The Client
- The Internal Consultant
- The External Consultant
- The Work Team
- The Organization

While you may have an initiative where there are more job roles or where you wish to break those listed above down to more basic parts, the five above must be included on your chart.

A2.4 A RACI Example

For our example, let's use the standard job steps included in the majority of change initiatives. Assume that, in addition to your role as internal consultant, the company has also decided to hire an external consultant. The tasks for the RACI Chart are:

1. The vision of the effort
2. Clarification of the assignment
3. Strategy development
4. Information gathering
5. Analysis and gap identification
6. Preparation of the recommendations
7. Presentation with defined next steps
8. Agreement—locking it down
9. Execution
10. Disengagement
11. Audit

RACI Chart	Roles				
	Client	Internal Consultant	External Consultant	Work Team	Organization
The vision of the effort					
Clarification of the assignment					
Strategy development					
Information gathering					
Analysis and gap identification					
Preparation of the recommendations					
Presentation with defined next steps					
Agreement					
Execution					
Disengagement					
Audit					

Figure A2-1 A Blank RACI Chart

With the job roles and tasks identified, the RACI Chart showing this information can be filled out, as shown in Figure A2-1.

Once this part of the effort has been completed, it is time to sit down with those who will be involved and build the actual chart. It is important that all parties are involved so that, when the chart is completed, everyone understands their roles and responsibilities. Although the client may not always be able to attend, you need to keep them updated and pass their feedback back to the group. In this way you will avoid finishing the work, only to have the client force you to rework it because they do not like the distribution of work. A completed example based on the job tasks identified is shown in Figure A2-2.

RACI Chart	Roles				
	Client	Internal Consultant	External Consultant	Work Team	Organization
The vision of the effort	A,R	R	R	R	I
Clarification of the assignment	A	R	R	I	
Strategy development	C	R	R	R	
Information gathering		C	C	A,R	I
Analysis and gap identification	I	C	C	R	
Preparation of the recommendations		C	C	A,R	
Presentation with defined next steps	A			R	
Agreement	A,R				
Execution	A			R	R
Disengagement	A	R	R	I	
Audit	A	R	R (maybe)	R	C,I

Figure A2-2 A Completed RACI Chart

APPENDIX 3

THE INTERNAL CONSULTANT'S WEB OF CHANGE

In Appendix 3, I have provided you with the internal consultant's web of change survey exactly as it is on the CD. This section is for those of you who wish to do the survey work manually. It also can be of value if you want to conduct the survey in a group setting and make copies to hand out. Once you have the group results, you can then transpose them to the electronic version.

Follow the directions in Chapter 16, which explains what to do with both the internal consultant's and the client's part of the survey. I have provided webs for both. I have also provided one for the creation of a composite which overlays your web with that of the client. This will enable you to see the performance gaps and take necessary corrective action.

The Composite Web

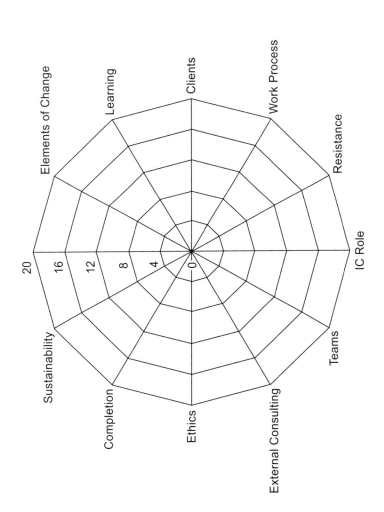

Enter the scores for each of the elements on this table and then use them to create you web.

The Internal Consultant's Web
Scores for the Elements

Element	Chapter	Score
Business Case	4	
Elements of Change	5	
Learning	6	
Clients	7	
Work Process	8	
Resistance	9	
IC Role	10	
Teams	11	
External Consulting	12	
Ethics	13	
Completion	14	
Sustainability	15	

Your Web

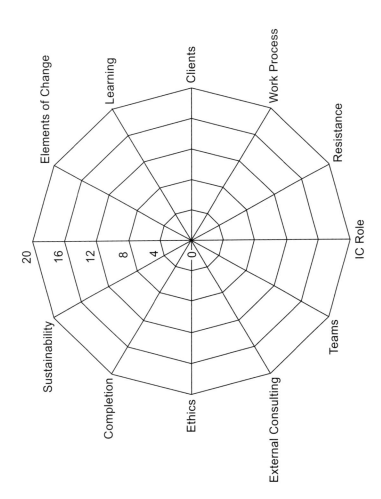

The Business Case (Chapter 4)	**Point**	**Score**
Total Score		

1   I recognize the need for a business case to justify the role of internal consultant as well as for many other strategic initiatives that need to be implemented.

A. Strongly agree	4
B. Agree	3
C. Neutral	2
D. Disagree	1
E. Strongly disagree	0

2   I understand the decision making process within my company well enough that I can proceed in the correct manner to maximize my chances for approval.

A. Strongly agree	4
B. Agree	3
C. Neutral	2
D. Disagree	1
E. Strongly disagree	0

3   I understand the outline of the business case model and can use it in the development of a business case.

A. Strongly agree	4
B. Agree	3
C. Neutral	2
D. Disagree	1
E. Strongly disagree	0

4   I understand and can develop details for each of the six components that may need to be included in the business case to show where value can be added.

A. Strongly agree	4
B. Agree	3
C. Neutral	2
D. Disagree	1
E. Strongly disagree	0

5   I have used the business case model (and enhanced it as I learned) as one of the tools in my internal consultant toolbox to promote strategic initiatives within my company.

A. Strongly agree	4
B. Agree	3
C. Neutral	2
D. Disagree	1
E. Strongly disagree	0

Elements of Change (Chapter 5)   Point   Score
Total Score

1   I understand the importance of addressing the Eight Elements of Change in any change initiative on which I am working.

A. Strongly agree	4
B. Agree	3
C. Neutral	2
D. Disagree	1
E. Strongly disagree	0

2   I understand the importance of addressing the Four Elements of Culture in any change initiative on which I am working.

A. Strongly agree	4
B. Agree	3
C. Neutral	2
D. Disagree	1
E. Strongly disagree	0

3   I understand the interdependence of the elements on each other and recognize the problems that failure to include them all will have on any change initiative.

A. Strongly agree	4
B. Agree	3
C. Neutral	2
D. Disagree	1
E. Strongly disagree	0

4   Not only do I understand the need to include the elements as part of the change process, but I take the time to address them on all change initiatives.

A. Strongly agree	4
B. Agree	3
C. Neutral	2
D. Disagree	1
E. Strongly disagree	0

5   I have a developed a detailed plan to educate others (especially my clients) of the need to address the elements of change within the process design and deployment.

A. Strongly agree	4
B. Agree	3
C. Neutral	2
D. Disagree	1
E. Strongly disagree	0

Group Learning (Chapter 6)	Point	Score
Total Score		

1   I fully understand my organization's level of maturity related to group learning. In areas where I recognize that it doesn't exist or isn't thriving, I try to promote it.

A. Strongly agree	4
B. Agree	3
C. Neutral	2
D. Disagree	1
E. Strongly disagree	0

2   I can see how group learning is a key component of successfully integrating the Eight Elements of Change into a work redesign effort. I work to promote this concept as part of my internal consultant role.

A. Strongly agree	4
B. Agree	3
C. Neutral	2
D. Disagree	1
E. Strongly disagree	0

3   I can see how group learning is a key component of successfully integrating the Four Elements of Culture into a work redesign effort. I work to promote this concept as part of my internal consultant role.

A. Strongly agree	4
B. Agree	3
C. Neutral	2
D. Disagree	1
E. Strongly disagree	0

4   I recognize whether my organization fits the learning or blaming model and consciously try to reinforce and support the former and eliminate the latter.

A. Strongly agree	4
B. Agree	3
C. Neutral	2
D. Disagree	1
E. Strongly disagree	0

5   I have consciously considered and developed a work strategy that encourages the group learning process in my internal consultant engagements.

A. Strongly agree	4
B. Agree	3
C. Neutral	2
D. Disagree	1
E. Strongly disagree	0

Clients (Chapter 7)	Point	Score
	Total Score	

1   I recognize that a client is needed for all internal consultant engagements if I wish to be part of a successful outcome.

A. Strongly agree	4
B. Agree	3
C. Neutral	2
D. Disagree	1
E. Strongly disagree	0

2   I understand how the definition of "client" is applied to working at the proper level within the organization and how not working at the correct level can impede adoption of the work.

A. Strongly agree	4
B. Agree	3
C. Neutral	2
D. Disagree	1
E. Strongly disagree	0

3   I understand what is to be expected from the client and work towards making certain that it is provided.

A. Strongly agree	4
B. Agree	3
C. Neutral	2
D. Disagree	1
E. Strongly disagree	0

4   I fully understand the client's expectations of me as the internal consultant and work hard to deliver on them.

A. Strongly agree	4
B. Agree	3
C. Neutral	2
D. Disagree	1
E. Strongly disagree	0

5   I understand that my ideas and suggestions may not be accepted and I do not take it personally. My job is to offer these ideas for consideration by my client.

A. Strongly agree	4
B. Agree	3
C. Neutral	2
D. Disagree	1
E. Strongly disagree	0

Work Process (Chapter 8)   Point   Score
Total Score

1   I understand and can explain the eleven steps of the internal consultant work process and how to apply them to the work that I do.

A. Strongly agree	4
B. Agree	3
C. Neutral	2
D. Disagree	1
E. Strongly disagree	0

2   Not only do I understand the steps, but also I apply them as part of the internal consulting work that I do.

A. Strongly agree	4
B. Agree	3
C. Neutral	2
D. Disagree	1
E. Strongly disagree	0

3   I understand the various techniques to acquire site "as is" information and know when to use the various strategies in executing this task.

A. Strongly agree	4
B. Agree	3
C. Neutral	2
D. Disagree	1
E. Strongly disagree	0

4   I recognize the need to apply the soft skills (Eight Elements of Change) and the Four Elements of Culture when performing gap analysis if I wish to be able to propose value added gap closure solutions.

A. Strongly agree	4
B. Agree	3
C. Neutral	2
D. Disagree	1
E. Strongly disagree	0

5   I understand and have (or will) apply the 10 steps required for a great presentation of the findings and recommended next steps.

A. Strongly agree	4
B. Agree	3
C. Neutral	2
D. Disagree	1
E. Strongly disagree	0

Resistance (Chapter 9)

Point   Score

Total Score

1 I am very aware of the various forms of resistance and know how to address them when they appear as part of the change process.

  A. Strongly agree   4
  B. Agree   3
  C. Neutral   2
  D. Disagree   1
  E. Strongly disagree   0

2 I can and have explained the forms of resistance to the change team so that they also understand what resistance looks like and can address it when they observe it.

  A. Strongly agree   4
  B. Agree   3
  C. Neutral   2
  D. Disagree   1
  E. Strongly disagree   0

3 I clearly recognize that resistance is not something to overcome, but rather something that needs to be understood and addressed. I have specific examples where I have addressed resistance properly.

  A. Strongly agree   4
  B. Agree   3
  C. Neutral   2
  D. Disagree   1
  E. Strongly disagree   0

4 I understand the concept of goal agreement and power balance, and know which technique to apply to support goal achievement across the organization.

  A. Strongly agree   4
  B. Agree   3
  C. Neutral   2
  D. Disagree   1
  E. Strongly disagree   0

5 I know and have applied Force Field Analysis in order to promote change and reduce the forces opposing it.

  A. Strongly agree   4
  B. Agree   3
  C. Neutral   2
  D. Disagree   1
  E. Strongly disagree   0

The Internal Consultant's Role (Chapter 10) Point   Score
Total Score

1 I understand the importance of the ion suffix action words in my role as an effective internal consultant.

A. Strongly agree	4
B. Agree	3
C. Neutral	2
D. Disagree	1
E. Strongly disagree	0

2 I work hard to include the actions described by these terms into the internal consultant work that I do.

A. Strongly agree	4
B. Agree	3
C. Neutral	2
D. Disagree	1
E. Strongly disagree	0

3 I recognize the areas I have for improvement and have developed corrective action plans.

A. Strongly agree	4
B. Agree	3
C. Neutral	2
D. Disagree	1
E. Strongly disagree	0

4 I make certain that my client understands all of the various actions required to deliver a successful change initiative.

A. Strongly agree	4
B. Agree	3
C. Neutral	2
D. Disagree	1
E. Strongly disagree	0

5 I have communicated all of the required actions, as described by the ion suffixes, to the work team. I recognize the importance of the need for the team to understand these actions so that they can be incorporated into the change process.

A. Strongly agree	4
B. Agree	3
C. Neutral	2
D. Disagree	1
E. Strongly disagree	0

Work Teams (Chapter 11)   **Point**   **Score**
                          Total Score

1   I understand the various components of the definition of team and have made sure that each was addressed when I help build or work with teams.
    A. Strongly agree   4
    B. Agree   3
    C. Neutral   2
    D. Disagree   1
    E. Strongly disagree   0

2   I understand the various types of teams and can support the building of team effectiveness as appropriate.
    A. Strongly agree   4
    B. Agree   3
    C. Neutral   2
    D. Disagree   1
    E. Strongly disagree   0

3   When helping to build teams, I make certain that the various membership requirements are addressed.
    A. Strongly agree   4
    B. Agree   3
    C. Neutral   2
    D. Disagree   1
    E. Strongly disagree   0

4   I understand my role as internal consultant in the team work process, guiding not leading, and have worked in this manner.
    A. Strongly agree   4
    B. Agree   3
    C. Neutral   2
    D. Disagree   1
    E. Strongly disagree   0

5   I recognize the need for inclusion of the Eight Elements of Change and the Four Elements of Culture in team development and success.
    A. Strongly agree   4
    B. Agree   3
    C. Neutral   2
    D. Disagree   1
    E. Strongly disagree   0

External Consultants (Chapter 12)

Total Score

Point   Score

1 When identifying an external consultant to support a change initiative, I have considered (or have others consider) how well they will fit with the work team as well as with my role as internal consultant.

A. Strongly agree	4
B. Agree	3
C. Neutral	2
D. Disagree	1
E. Strongly disagree	0

2 I am aware of the control / involvement issues as depicted in Figure 12-1 and work to promote Quadrant 4 interaction.

A. Strongly agree	4
B. Agree	3
C. Neutral	2
D. Disagree	1
E. Strongly disagree	0

3 I am aware of the value that external and internal consultants each bring to the effort and work to optimize the value each can deliver.

A. Strongly agree	4
B. Agree	3
C. Neutral	2
D. Disagree	1
E. Strongly disagree	0

4 I understand the RACI process and have utilized (or will utilize) it to clearly establish roles and responsibilities of the various individuals or groups involved in a change initiative.

A. Strongly agree	4
B. Agree	3
C. Neutral	2
D. Disagree	1
E. Strongly disagree	0

5 I understand how both clients and consultants can hinder progress of an initiative and work to mitigate these problems.

A. Strongly agree	4
B. Agree	3
C. Neutral	2
D. Disagree	1
E. Strongly disagree	0

Ethics (Chapter 13)

Point   **Score**
Total Score

1   As an internal consultant, I practice the concept of "do not harm" in the work that I perform.

       A. Strongly agree   4
       B. Agree   3
       C. Neutral   2
       D. Disagree   1
       E. Strongly disagree   0

2   I understand the three reasons for unethical choices and pay close attention to the choices I make to make certain I don't fall into these traps.

       A. Strongly agree   4
       B. Agree   3
       C. Neutral   2
       D. Disagree   1
       E. Strongly disagree   0

3   I pay close attention to potential conflicting ethical situations and when they arise I work to create a balance and resolve them favorably. I strive for Quad 4 ethical performance.

       A. Strongly agree   4
       B. Agree   3
       C. Neutral   2
       D. Disagree   1
       E. Strongly disagree   0

4   I have reviewed the 20 guidelines and recognize the ones that are my strengths. I try to model these behaviors for those with whom I get involved as an internal consultant.

       A. Strongly agree   4
       B. Agree   3
       C. Neutral   2
       D. Disagree   1
       E. Strongly disagree   0

5   I recognize my weaknesses and have developed corrective action plans to improve.

       A. Strongly agree   4
       B. Agree   3
       C. Neutral   2
       D. Disagree   1
       E. Strongly disagree   0

Completion (Chapter 14)   Point   Score
                           Total Score

1   I have built (early on) an exit plan into the work initiatives in which I am engaged.
      A. Strongly agree   4
      B. Agree   3
      C. Neutral   2
      D. Disagree   1
      E. Strongly disagree   0

2   Part of my completion effort includes assuring that the site takes full ownership of what has been developed.
      A. Strongly agree   4
      B. Agree   3
      C. Neutral   2
      D. Disagree   1
      E. Strongly disagree   0

3   I know what signs to look for that indicate a potential failure of an initiative to complete.
      A. Strongly agree   4
      B. Agree   3
      C. Neutral   2
      D. Disagree   1
      E. Strongly disagree   0

4   Once I recognize the potential failure to complete signs, I know how to take corrective action to mitigate or eliminate the problem.
      A. Strongly agree   4
      B. Agree   3
      C. Neutral   2
      D. Disagree   1
      E. Strongly disagree   0

5   I have created (or will create) a contract with any initiative team with which I am to work so that they understand what I will and will not deliver in order to support the effort and promote my exit strategy.
      A. Strongly agree   4
      B. Agree   3
      C. Neutral   2
      D. Disagree   1
      E. Strongly disagree   0

Readiness and Sustainability (Chapter 15)   Point   Score
Total Score

1   I understand the need for readiness and sustainability to be a part of each change initiative. I also recognize my role to make sure that this takes place.

A. Strongly agree	4
B. Agree	3
C. Neutral	2
D. Disagree	1
E. Strongly disagree	0

2   I have applied (or will apply) the concepts of readiness and sustainability to all initiatives in which I am involved.

A. Strongly agree	4
B. Agree	3
C. Neutral	2
D. Disagree	1
E. Strongly disagree	0

3   I understand and will apply the Goal Achievement Model concept to the development of change readiness plans.

A. Strongly agree	4
B. Agree	3
C. Neutral	2
D. Disagree	1
E. Strongly disagree	0

4   I recognize how the Four Elements of Culture need to be crafted into the sustainability plan and have followed (or will follow) this process.

A. Strongly agree	4
B. Agree	3
C. Neutral	2
D. Disagree	1
E. Strongly disagree	0

5   I will present the need for readiness and sustainability to my clients and to the work teams with whom I become involved so that I can make sure these concepts are incorporated into each change initiative.

A. Strongly agree	4
B. Agree	3
C. Neutral	2
D. Disagree	1
E. Strongly disagree	0

The Client's Web

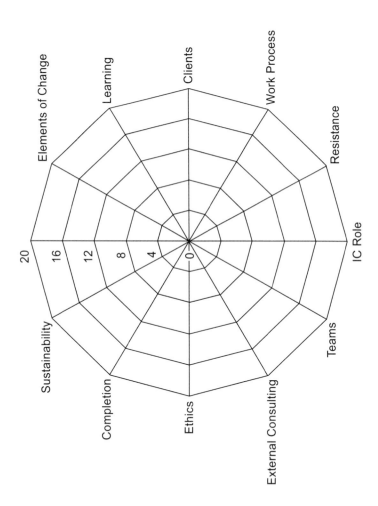

Client Survey Scores

Name >>>								Average Score per Statement	Average Score per Category

Enter the scores from the client survey forms below. All elements of the web are represented as well as each of the questions.

Business Case (Chapter 4)

1									
2									
3									
4									
5									

Elements of Change (Chapter 5)

1									
2									
3									
4									
5									

Learning (Chapter 6)

1									
2									
3									
4									
5									

Clients (Chapter 7)

1									
2									
3									
4									
5									

Work Process (Chapter 8)

1									
2									
3									
4									
5									

Resistance (Chapter 9)

1									
2									
3									
4									
5									

Client Survey Scores

Name >>>							Average Score per Statement	Average Score per Category

Enter the scores from the client survey forms below. All elements of the web are represented as well as each of the questions.

IC Role Chapter 10)

1								
2								
3								
4								
5								

Teams (Chapter 11)

1								
2								
3								
4								
5								

Ext Consulting (Chapter 12)

1								
2								
3								
4								
5								

Ethics (Chapter 13)

1								
2								
3								
4								
5								

Completion (Chapter 14)

1								
2								
3								
4								
5								

Sustainability (Chapter 15)

1								
2								
3								
4								
5								

Client Feedback Survey

Name:

Business Case (Chapter 4)

		Strongly Agree	Agree	Neutral	Disagree	Strongly Disagree	Comments for scores of 0 or 1
1	The internal consultant recognizes the need to employ the business case model.	4	3	2	1	0	
2	The internal consultant understands the decision making process to proceed in the correct manner to maximize the chance of approval.	4	3	2	1	0	
3	The internal consultant understands the outline of the business case model and can use it in the development of a business case.	4	3	2	1	0	
4	The internal consultant understands and can develop details for each of the six value adding components that may need to be included in the business case.	4	3	2	1	0	
5	The internal consultant uses the business case model as one of the tools in their internal consultant toolbox to promote strategic initiatives.	4	3	2	1	0	

Client Feedback Survey

Name:

Elements of Change (Chapter 5)	Strongly Agree	Agree	Neutral	Disagree	Strongly Disagree	Comments for scores of 0 or 1
1	The internal consultant has a very clear understanding of all of the elements of change and how they need to be included in any change initiative.	4	3	2	1	0	
2	The internal consultant has clearly communicated and convinced me of the need for inclusion of the elements of change in the change initiative design and deployment.	4	3	2	1	0	
3	The internal consultant has addressed the Eight Elements of Change in the development and deployment of the change initiative.	4	3	2	1	0	
4	The internal consultant has addressed the Four Elements of Culture in the development and deployment of the change initiative.	4	3	2	1	0	
5	The internal consultant has addressed the interdependency of the elements and made certain that none were missing in developing the change process.	4	3	2	1	0	

Client Feedback Survey

Name:							
Group Learning (Chapter 6)	Strongly Agree	Agree	Neutral	Disagree	Strongly Disagree	Comments for scores of 0 or 1
1	The internal consultant understands the value of group learning as it is applied to organizational change.	4	3	2	1	0	
2	The internal consultant had promoted double loop learning wherever and whenever possible to help us reevaluate our goals; just not get better at the same things we have always done.	4	3	2	1	0	
3	The internal consultant made certain to include group learning as part of the process both in the development and deployment phases.	4	3	2	1	0	
4	Where group learning was not in evidence the internal consultant worked to integrate this concept into the work being conducted.	4	3	2	1	0	
5	The internal consultant understands the concept of a blaming organization, has pointed out where this philosophy exists and has worked with the organization to try to change it.	4	3	2	1	0	

Client Feedback Survey

Name:

Clients (Chapter 7)	Strongly Agree	Agree	Neutral	Disagree	Strongly Disagree	Comments for scores of 0 or 1
1. The internal consultant is fully aware and can explain why a client is needed for a work initiative to be successful.	4	3	2	1	0	
2. The internal consultant has completely explained to those involved the danger of not having a client as a key player in any work initiative.	4	3	2	1	0	
3. The internal consultant has made my role as client completely clear so that I can perform it.	4	3	2	1	0	
4. When I do not perform the role which I am expected to, the internal consultant keeps me aware of things I should be doing in support of the initiative but am not.	4	3	2	1	0	
5. The internal consultant has built a high level of credibility with me and with the organization. I respect the internal consultant's opinion and take action on their recommendations.	4	3	2	1	0	

Client Feedback Survey

Name:							
The Work Process (Chapter 8)	Strongly Agree	Agree	Neutral	Disagree	Strongly Disagree		Comments for scores of 0 or 1
1	The internal consultant understands and has explained the eleven steps of the internal consultant work process to me and I understand the process.	4	3	2	1	0	
2	The internal consultant has explained how they plan to apply the eleven steps to the work and based on this explanation I can see how this will add value as well as help the organization.	4	3	2	1	0	
3	The internal consultant has explained the various techniques to acquire site "as is" information and we have agreed on the strategy to be employed. I feel comfortable that this approach will identify our strengths as well as areas for improvement.	4	3	2	1	0	
4	Based on the internal consultant's explanation I recognize the need to apply the soft skills (Eight Elements of Change) and the Four Elements of Culture if I wish the change initiative to be successful.	4	3	2	1	0	
5	I understand the need for a pre-presentation of the internal consultant findings so that I can make certain I fully understand the material and can offer "visible" support.	4	3	2	1	0	

Client Feedback Survey

Name:

Resistance (Chapter 9)

		Strongly Agree	Agree	Neutral	Disagree	Strongly Disagree	Comments for scores of 0 or 1
1	The internal consultant is aware of the various forms of resistance and knows how to address them when they appear as part of the change process.	4	3	2	1	0	
2	The internal consultant can (and has) explained the forms of resistance to the change team so that they understand what resistance looks like and can address it when it is observed.	4	3	2	1	0	
3	The internal consultant understands that resistance is not something to overcome but rather something that needs to be understood and addressed. There have been visible examples of this in the conduct of the change process.	4	3	2	1	0	
4	The internal consultant understands the concept of goal agreement and power balance and knows which technique to apply to support goal achievement across the organization.	4	3	2	1	0	
5	The internal consultant knows how to apply Force Field Analysis in order to promote change and reduce the forces opposing it.	4	3	2	1	0	

Client Feedback Survey

Name:

Internal Consultant's Role (Chapter 10)	Strongly Agree	Agree	Neutral	Disagree	Strongly Disagree	Comments for scores of 0 or 1
1	The internal consultant has clearly communicated information about the *ion* suffixes to me so that I understand what to expect as part of the effort and the deliverables.	4	3	2	1	0	
2	It is clear to me that the internal consultant understands the *ion* suffixes and how they relate to the internal consultant's role.	4	3	2	1	0	
3	The internal consultant includes the action represented by these terms in the work that they perform.	4	3	2	1	0	
4	I have spoken with the internal consultant and pointed out areas of strength and areas for improvement as a part of the initiative feedback that I provide.	4	3	2	1	0	
5	The internal consultant has communicated the *ion* suffix information to the work team and worked with them to assure these actions are addressed as part of the change initiative.	4	3	2	1	0	

Client Feedback Survey

Name:							
Work Teams (Chapter 11)	Strongly Agree	Agree	Neutral	Disagree	Strongly Disagree	Comments for scores of 0 or 1
1	The internal consultant clearly understands the various components of the definition of team and has made sure that each was addressed when they worked with teams.	4	3	2	1	0	
2	The internal consultant understands the various types of teams and can support the building of team effectiveness as appropriate.	4	3	2	1	0	
3	The internal consultant has worked closely with the management team to make certain that the various team membership requirements were addressed.	4	3	2	1	0	
4	As I have observed the internal consultant functioning within the team, I recognize that they understand their role in the process.	4	3	2	1	0	
5	The internal consultant has included work with the Eight Elements of Change and the Four Elements of Culture in team's development.	4	3	2	1	0	

Client Feedback Survey

Name:							
External Consultants (Chapter 12)	Strongly Agree	Agree	Neutral	Disagree	Strongly Disagree	Comments for scores of 0 or 1
1	The internal consultant has worked with senior management when identifying an external consultant so that there is a good fit between them, the internal consultant and the work team.	4	3	2	1	0	
2	The internal consultant is aware of the control / involvement issues as depicted in figure 12-1 and works to promote Quadrant 4 interaction.	4	3	2	1	0	
3	The internal consultant is aware of the value that external and internal consultants each bring to the effort and works to optimize the value each can deliver.	4	3	2	1	0	
4	The internal consultant understands the RACI process and has utilized it to clearly establish roles and responsibilities of the various individuals or groups involved in a change initiative.	4	3	2	1	0	
5	The internal consultant understands how both clients and consultants can hinder progress of an initiative and works to mitigate these problems.	4	3	2	1	0	

Client Feedback Survey

Name:

Ethics (Chapter 13)	Strongly Agree	Agree	Neutral	Disagree	Strongly Disagree	Comments for scores of 0 or 1
1	I have observed the internal consultant and believe that they practice the concept of "do no harm" in the work that they perform.	4	3	2	1	0	
2	I understand the three reasons for unethical choices and have observed the internal consultant working to avoid falling into the traps they present.	4	3	2	1	0	
3	The internal consultant pays close attention to potential conflicting ethical situations and works to favorably resolve them.	4	3	2	1	0	
4	I have discussed the 20 guidelines with the internal consultant. I believe that they recognize their strengths and try to model these behaviors.	4	3	2	1	0	
5	I know that the internal consultant has identified their weaknesses related to the guidelines and has developed corrective action plans to improve themselves.	4	3	2	1	0	

Client Feedback Survey

Name:

Completion (Chapter 14)	Strongly Agree	Agree	Neutral	Disagree	Strongly Disagree	Comments for scores of 0 or 1
1	The internal consultant has made me aware early in the initiative the need for an exit strategy and I understand why.	4	3	2	1	0	
2	In my dealings with the internal consultant, they have made it clear that one of their key completion roles is to assure that the site takes full ownership of what has been developed.	4	3	2	1	0	
3	The internal consultant pays careful attention to the signs that indicate a potential failure of an initiative to complete. They have explained the rational for this to me.	4	3	2	1	0	
4	The internal consultant has made me aware when signs indicating a potential failure to complete are present and have provided corrective action plans.	4	3	2	1	0	
5	The internal consultant has created a contract with the change team so that they understand what they need to do to take on full ownership.	4	3	2	1	0	

Client Feedback Survey

Name:

Sustainability (Chapter 15)	Strongly Agree	Agree	Neutral	Disagree	Strongly Disagree	Comments for scores of 0 or 1
1	The internal consultant understands the need for readiness and sustainability to be a part of each change initiative. They have taken responsibility to make sure that this takes place.	4	3	2	1	0	
2	The internal consultant has applied the concepts of readiness and sustainability to the initiatives and activities that I have observed.	4	3	2	1	0	
3	The internal consultant understands and has applied the Goal Achievement Model concept to the development of change readiness plans.	4	3	2	1	0	
4	The internal consultant recognizes that the Four Elements of Culture need to be crafted into the sustainability plan and has followed this process.	4	3	2	1	0	
5	The internal consultant has presented the need for readiness and sustainability to me and to the work team that they are supporting. They have worked hard to make certain that these concepts are incorporated into the change initiative.	4	3	2	1	0	

BIBILOGRAPHY

Block, Peter. *Flawless Consulting*. Penton Overseas, Inc., and Audio Scholar, Inc., 2001.

Chowdhury, Subir. *The Power of Six Sigma*. Chicago: Dearborn Trade Publishing, 2001.

Deal, Terrence E. and Allen A. Kennedy. *Corporate Cultures*. New York: Addison Wesley, 1982.

Drucker, Peter F. *Management: Tasks, Responsibilities, Practices*. New York: Harper & Row, 1974.

Eldred, John. "Mastering Organizational Politics and Power." Masters course presented at the University of Pennsylvania. Philadelphia, PA., 1994.

Hinckley, Jr., Stanley R. "Managing Internal and External OD Consultants." *Consultation Skills Reading*. Rosslyn Station: NTL Institute, 1984.

Machiavelli, Niccolo. *The Prince*. New York: Bantam Books, 1966.

Maxwell, John. *Business Ethics—Why do some people make unethical choices?* Retrieved October 20, 2006 from Beginners Guide Web site: http://beginners-giude.com

Nelms, Robert. *What You Can Learn from Things that Go Wrong—A Guidebook to the Root Causes of Failure*. Virginia: Failsafe Network [year?]

Schaffer, Robert H. *High Impact Consulting*. San Francisco: Jossey-Bass, 1997.

Schein, Edgar H. *Organizational Culture and Leadership, 2nd Ed*. San Francisco: Jossey Bass, 1992.

Schein, Edgar H. *The Corporate Culture Survival Guide*. San Francisco: Jossey Bass, 1999.

Swartz, Donald H. "Similarities and Differences of Internal and External

Consultants." *Consultation Skills Reading*. Rosslyn Station: NTL Institute, 1984.

Thomas, Stephen J. *Successfully Managing Change in Organizations: A Users Guide*.
New York: Industrial Press, 2001.

Thomas, Stephen J. *Improving Maintenance and Reliability Through Cultural Change*. New York: Industrial Press, 2005.

Weiss, Alan, PhD. *Organizational Consulting*. Hoboken: John Wiley &Sons, 2003.

LIST OF FIGURES

LIST OF TABLES

INDEX